SpringerBriefs in Molecular Science

Biometals

More information about this series at http://www.springer.com/series/10046

Anil K. Suresh

Co-Relating Metallic Nanoparticle Characteristics and Bacterial Toxicity

 Springer

Anil K. Suresh
Department of Biotechnology,
   School of Life Sciences
Pondicherry University
Pondicherry
India

ISSN 2191-5407   ISSN 2191-5415   (electronic)
SpringerBriefs in Molecular Science
ISSN 2212-9901
SpringerBriefs in Biometals
ISBN 978-3-319-16795-4   ISBN 978-3-319-16796-1 (eBook)
DOI 10.1007/978-3-319-16796-1

Library of Congress Control Number: 2015935412

Springer Cham Heidelberg New York Dordrecht London

Printed on acid-free paper

Springer International Publishing AG Switzerland is part of Springer Science+Business Media
(www.springer.com)

Preface

Nanotechnology is an emerging and dominating area that includes a novel class of nanomaterials that are being considered for various applications including electronics, smart devices, sensors, and biomedicine. Nanotechnology has huge potential in all fields of science and technology, because of their morphology-dependent unique physical, chemical, electronic, catalytic, and biological properties. Nanotechnology has already begun to have a huge impact on various perspectives of beneficiary aspects, and is drastically revolutionizing the industries and pharmaceutical companies with great emphasis on human health, environment safety, and sustainability. Nanotechnology has already begun to pervade several aspects of our day-to-day life, and researchers are revisiting several useful aspects with a nanoperspective for a better lifestyle. This phenomenon is likely revolutionizing medical sciences, and many chemotherapeutics are being reconsidered for possible improvements using engineered nanoparticles.

This book will emphasize issues related to the safe use of nanoparticles, keeping in mind the biotic environment. Nanoparticles in various forms are tremendously being implemented in several products as additives and therefore are getting released into the environment as pollutants. Interactions between the nanoparticles and microorganisms in the environment are unavoidable, but the pandemic consequences of such interactions are beginning to be investigated. This brief book will illustrate on how naturally occurring microorganisms and man-made nanoparticles interact, and the consequences of such interaction, using suitable examples from our studies published in several peer-reviewed international journals. This book will not only be helpful for the scientific and industrial community but will also attract wide attention of students and researchers in different areas of sciences such as microbiology, biotechnology, nanotechnology, toxicology, materials science, biomedical engineering, and cell and molecular biology.

The several objectives of this brief book are to introduce nanobiotechnology along with "nanotoxicology" aspects, and make the readers aware of the potential interactions of engineered nanoparticles with microorganisms. Impacts of toxic metal and metal oxide nanoparticles such as silver and zinc oxide on the growth and viability of several bacteria are presented. Differences in the bacteria–nanoparticle interactions using different forms of nanoparticles, nanoparticles synthesis methodologies are described with emphasis on the influence of surface coatings. The use of various analytical and physical characterization techniques that are often used to analyze nanoparticle bacteria interactions are outlined. Mechanistic insights into the relationship between the bacterial growth inhibition, reactive oxygen species generation, and up- and/or downregulation of transcriptional stress-responsive genes are also discussed. Finally, how advanced and emerging imaging techniques such as transmission electron and atomic force microscopes can be made use to assess their interactions are discussed, which will have impacts toward better understanding on the overall microbial–nanoparticle interactions.

The book contains four chapters; Chap. 1 includes a general introduction to various nanoparticles that are considered lethal to microbial cells (bacteria, virus, and fungus) with emphasis on metal and metal oxide nanoparticles. Interactions of various nanoparticles with microbes along with the influence of additives in the form of solvents, surfactants, and capping materials are described using suitable examples. Various proposed mechanisms by which the nanoparticles induce toxicity and the bacterial stress response toward nanoparticles are described using multiple examples. Chapter 2 describes the commonly used laboratory experimental, analytical, and physical characterization techniques to evaluate and determine the toxicity of nanoparticles toward different microorganisms. Comparative assessments on the differences between these procedures are described correlating to nanoparticle properties. The role of multianalytical assays and techniques used for understanding the interactions of nanoparticles with microbial cell systems are presented. The growth and viability on the bacteria relative to nanoparticles size, growth media, and dosage are discussed. Details of the bactericidal impacts assessed using multiple assays such as minimum inhibitory concentration, colony-forming units, disk diffusion tests, and live/dead assays are provided. Discussions of advanced tools such as inductively coupled plasma–mass spectroscopy, scanning electron microscopy, transmission electron microscopy, dark-field microscopy, and atomic force microscopy that are used to understand the response mechanism of the bacteria are outlined. Chapter 3 describes the bactericidal properties of zinc oxide nanoparticles and the detailed mechanistic of their interaction with regard to the bacterial viability, reactive oxygen species (ROS) generation, and surface alterations on the bacterial cell. The relation between the nanoparticle and bacteria interaction with respect to transcriptional genome stress profiling is described. Analysis on the various up- and downregulation of genes based on microarrays to evaluate the bacterial genetic response mechanism is described. Chapter 4 describes the influence of various surface coatings of nanoparticles in dictating bactericidal toxicity. Various surface stabilizing agents often used to synthesize nanoparticles, along with their roles with respect to particles size and shape control, particles over all charge, particles stability, interaction abilities with the biomaterials, and/or cells, etc., are discussed. How engineered nanoparticles are incorporated with various surface coatings during their synthesis, along with details of the various physical characterization techniques including zeta potential,

Fourier transform infrared spectroscopy, and X-ray photoelectron spectroscopy are described. Finally, comparative studies on the effects of various surface-coated nanoparticles on the toxicity of bacteria are discussed.

I am pleased that I have been invited to contribute this second brief book published by Springer within the Springer Briefs in Biometals series by Prof. Larry Barton, edited by Dr. Sonia Ojo. I am glad to submit this book and I hope that you will enjoy reading it more than I did while writing.

Puducherry   Anil K. Suresh
December 2014

Acknowledgments

Much to my extreme delight, I would like to evince my gratitude and indebtedness to my beloved Ph.D. mentor late. Dr. M.I. Khan, a truly remarkable scientist who introduced me to this fascinating realm of nanobioscience and nanobiotechnology. His invaluable guidance, constant inspiration, and unending support have always been contagious and motivational throughout my Ph.D. pursuit. His scientific temperament, innovative approach, dedication toward his profession and his down-to-earth and helping nature has inspired me highly. Although this eulogy does not give him justice, I preserve an everlasting gratitude for him.

I heartily thank Prof. Jay Nadeau at McGill University, Canada, and Prof. Yves-Alain Peter at Ecole Polytechnique in Montreal, Canada, for their mentorship during my first postdoctoral training. Special thanks to them for giving me the opportunity to let me explore my research expertise on nanomaterials for further implementation in cell imaging and drug delivery applications. They often used to organize family get-together and fun-filled extracurricular activities (kayaking, canoeing, skiing, rock climbing, and ropes courses). It was a very friendly environment.

I wish to express my sincere gratitude and heartfelt thanks to Dr. Mitchel Doktycz and Dr. Dale Pelletier at Oak Ridge National Laboratory in Oak ridge, USA. With them, I pursued my second postdoctoral training. I am grateful for their mentorship, motivation, subtle guidance, fruitful discussions, never-ending support, and constant help. The trust and freedom they gave me to implement my own research ideas have been crucial to achieve this feat. Working with them has always made me feel relaxed and has enabled me to progress in a friendly and pleasant environment. I will never been able to thank them enough. They are godfathers to me.

I would also like to use this opportunity to express my sincere thanks to Prof. Jacob Berlin, with whom I worked as a Staff Scientist at the Department of Molecular Medicine, Beckman Research Institute at City of Hope, USA. Due thanks to him for introducing me to my dream area of research: cancer therapeutics and clinical medicine. It was a great pleasure working with him; his generosity and kind nature inspire me the most. I will never be able to thank him enough for his remarkable forbearing at times and all his support.

I sincerely acknowledge and I am very much thankful to my research collaborators and colleagues Dr. Sudhakar, Dr. Arun Kumar, Dr. Arul, Dr. Arumugam, Dr. Sakthivel, Dr. Prashant, Dr. Sharma, Dr. Hanna, and Dr. Latha, at the Department of Biotechnology, Pondicherry University, for their valuable help whenever necessary and constant support.

It gives me immense pleasure to thank my mom, dad, and my lovely brothers Sunil and Vinil, for their love, unfailing support, tremendous patience, trust, and encouragement shown in their own special ways during every step of my life. They have been a constant source of strength and inspiration for me. I also thank my sisters-in-law, Sujatha and Jyotshna, for their constant support and trust.

I would also like to thank my wife, Arundhati, for her care, understanding, and love.

I am ever grateful to Almighty God, the Creator and the Guardian, and to whom I owe my very existence; because of his blessings, wisdom and perseverance that he has been bestowed upon me at all times. I bow to the divine strength and hope that his blessings will dwell throughout my life.

Last but not least, thanks to my lovely angel daughter Akanksha, whose cute smiles and love soothe the pain I experience while achieving every feat of my life.

Contents

Chapter 1
An Overview on Toxic Nanoparticles and Their Interactions with Microbial Cells

Prakash Gajapathi, Meyappan Vadivel, Anand Thirunavukarsou, Sudhakar Baluchamy and Anil K. Suresh

Abstract Engineered nanostructures include novel class of materials that are gaining tremendous recognition to pursuit in diverse disciplines including biomedicine, bioengineering, and electronics. Enormous applications of engineered nanoparticles in the existing and emerging technologies are leading to their large-scale production as well as environmental release. Therefore, to ensure proper uses of nanoparticles, it is important to understand the interactions of nanoparticles with biotic cells. Depending on the type of nanoparticle and its intrinsic properties, nanoparticles are considered either lethal or non-hazardous. Additionally, the size, shape, composition, and surface properties can significantly govern the performance of nanoparticles and might similarly influence their interaction with bacterial cells. This chapter will focus on overviewing the literature on the possible modes of interaction of engineered nanoparticles with the bacteria and along with the proposed mechanistic insights.

Keywords Nanoparticles · Nanotoxicity · Microbicidal · Lethality · Bacteria interactions

1.1 Microbial Toxicity of Nanoparticles: An Introduction

Nanoparticles are important class of scientific materials that are being evaluated for various biotechnological, pharmacological, and pure technological applications. Nanoparticles are the particles with dimensions at the nanometer scale, less than 100 nm [1]. At this length scale and depending on their form, nanoparticles possess unique properties such as large surface area, high catalytic activity, cohesive energy, chemical reactivity along with physico-chemical, optoelectronic, and biological properties that are different when compared to that of their respective bulk counterparts [2–5]. These unique properties attract wide attention of researchers for the development of better functional

© The Author(s) 2015
A.K. Suresh, *Co-Relating Metallic Nanoparticle Characteristics and Bacterial Toxicity*, SpringerBriefs in Biometals,
DOI 10.1007/978-3-319-16796-1_1

Table 1.1 List of nanoparticles commonly used for various consumer products and biomedical applications

Nanoparticle type	Product
Silver, zinc oxide, and copper oxide	Antimicrobial agents, antimicrobial paints, water purifications, textile, and medical devices
Cerium oxide	Automobiles exhaust
Nickel and lithium	Batteries
Titanium dioxide, copper oxide, and zinc oxide	Paints, ceramics, sunscreen, cosmetics, and catalysts
Gold, platinum and palladium	Sensors, catalysts, and fuel cells
Carbon nanotubes and fullerenes	Electronics, lubricants, and cosmetics
Iron oxide	MRI contrast agents, detection, magnetic separations, and medicinal devices

systems and smart devices, and products of nanoscience and nanotechnology are beginning to pervade several aspects of human life [6]. Engineered nanoparticles are becoming integral to smart technologies such as computer processors, advanced functional materials as well as consumer products such as cosmetics, sunscreens, automobiles, water purification systems, and textile [7, 8]. Several forms of nanoparticles are also used in biomedicine and therapeutic applications due to their potent bactericidal and fungicidal properties (Ag, ZnO, CuO, and TiO_2), and theranostic properties (Au and Fe_3O_4), and are widely implemented in medicinal products and devices, food packaging, and healthcare and household products [6].

The toxicities of engineered nanoparticles have all been reported for various prokaryotic and eukaryotic cell systems, whole organisms, and even plants [6]. Major forms of nanoparticles that are so far reported to have toxic effects include the fullerenes, carbon nanotubes, metal and metal oxide nanoparticles, quantum dots, metal complexes, and organic polymers. A summary of the common nanoparticles that are used for various consumer products as well as biomedical purposes is given in Table 1.1. Of the broad nanomaterial forms, metal nanoparticles represent one of the largest and most widely used classes. Therefore, it becomes highly imperative to be able to understand their potential interactions, fate and transport with environmental biotic systems, and human health [6].

The present chapter attempts to summarize the emerging toxicity aspects of nanomaterials with respect to prokaryotic cells. Firstly, a brief overview on the various nanoparticles and their potent bactericidal effect will be provided. Then, the influence of major contributing factors such as the form or type, synthesis methods employed, size and shape distributions, surface coatings, and aggregation potential in dictating the bacterial toxicity will be discussed with focus on metallic nanoparticles. Finally, the various modes of mechanisms that are so far proposed to ascertain nanoparticle bacterial interactions will be outlined.

1.2 Bacterial Toxicity of Nanoparticles

Microorganisms are widely recognized for their ability to adapt to their local environment and to exploit available energy resources in the existing forms [9, 10]. Bacteria pervading major parts of the biosphere such as air, water, and soil, within deep rocks and earth's crust, are considered critical for the formation as well as proper functioning of the normal abiotic as well as biotic ecosystem [11]. Bacteria act as decomposers and are considered crucial for several natural life cycles such as nitrogen and carbon recycling, nitrogen and ammonia fixation for the balance of ecological flora, fauna, and biotic environment. Apart from natural processes, bacteria have also been exploited for human benefit in the form of biotechnology in several applications such as food and beverages, decontamination, fermentation, production of antibiotics, bioremediation, wastewater treatments, pollution control, and mining of ores [12].

Engineered nanomaterials that are profoundly used for various industrial, environmental, and research applications and that are generated as by-products and/or accidentally released (e.g., coal ash, oil spills) are eventually getting exposed into the environment. These nanopollutants if effect soil bacteria would indirectly affect soil quality, sustainability, and ecosystem as a whole, including humans. Current understanding on the environmental effects of nanoparticle exposure lags significantly behind over the development and use of engineered nanoparticle-based technologies. When it comes to the interaction of engineered nanoparticles with bacterial cells, various efforts performed by several investigators have led to seemingly different contradictory assessments. Further complicating toxicity interpretations are the effects of the synthesis methodology; various manufacturing processes may incorporate additives, detergents, and solvent chemicals that are not completely eliminated from the final product. For instance, C60 that was initially deemed to be toxic [13], in later studies, indicated that remnants of tetrahydrofuran used during the synthesis of C60 were responsible for the toxicity [5]. Thus, the apparent biological properties of nanomaterials may depend in part on other constituents present in the formulation. Finally, not all nanoparticles are same and the properties of nanoparticles depend on their type or form in other words the primary construct. Bactericidal toxicity on some of the commonly used nanomaterials along with their bactericidal mechanism is discussed in the following sections.

1.2.1 Toxicity Induced by Fullerenes and Carbon Nanotubes

A fullerene can be any molecule that is composed of entirely a carbonaceous material, either in the form of a hollow spheres or ellipsoids or tubes. Structurally, fullerene is a form of graphite and is widely implemented in various medicinal and electronic applications, primarily, due to their heat resistance, superconductivity, and catalytic properties. As mentioned earlier, toxicity issues with respect to fullerenes are quiet controversial, for instance, C60 initially was suggested to be harmful [5, 13] but in later studies showed neutral biological effects [14]. However, studies by Fortner et al., while evaluating the effects of C60 aggregates on two common soil bacteria *E. coli* and *B. subtilis,* suggested that C60 exposure at relatively low concentrations inhibited the bacterial growth and decreased the respiration rates. In another investigation, when fullerenes were dispersed in water suspensions, their properties completely changed from being less toxic to highly bactericidal against *B. subtilis* [15]. Contrarily, it has also been suggested that the water-dispersed fullerenes exerts non-ROS-mediated oxidative stress toward bacteria, with evidence of protein oxidation, changes in cell membrane potential, and interruption of cellular respiration [16]. Similarly, water-dispersed polyvinylpyrrolidone, gamma-cyclodextrin, and fullerenols were found to be toxic to six different kinds of bacteria [17]. The toxicity of C60 has been attributed to its ability to bind and deform DNA stands and interference with DNA repair mechanisms. Likewise, carbon nanotubes (CNTs) have also been shown to exhibit bactericidal properties. It is speculated that the mechanistic on the bacterial toxicity of single-walled CNTs is due to the direct damage of the cell membrane, whereas multi-walled CNTs induce oxidative stress-mediated bactericidal activity [18]. Carbon-based nanomaterials have also been found to be inactive against *E. coli, S. epidermis,* beneficial soil microbes like *P. aeruginosa, B. subtilis* as well as diverse microbial communities of river and wastewater effluent [19].

1.2.2 Toxicity Induced by Metal Nanoparticles

Engineered metallic core/core–shell nanoparticles are the most widely used nanomaterials in biology and medicine. Bactericidal activity for several forms of metallic nanoparticles including Ag, Au, SiO_2, Al_2O_3, Fe_2O_4, TiO_2, ZnO, CdS, CdSe, CdTe, CdSe-ZnS, etc. have been reported and raises serious environmental concerns [5]. Differences in the reported microbial toxicities can also be ascribed to several additional factors [4, 5]. For example, the source of the materials used for their synthesis, form or type, size and shape distributions, surface charge, the presence of stabilizing or capping agents (discussed in detail in Chap. 4), the chemical and physical properties of the nanoparticles, particle behavior, particle aggregation phenomenon, the procedures employed to evaluate toxicity, the reaction medium, and how well the particles are characterized. Some of the important contributors that play a vital role in nanoparticle–microbial interaction are described below.

1.2.2.1 Influence of Nanoparticle Form or Type

Not all nanoparticles are same; the overall properties of nanoparticle primarily depend on their form or type, along with other characteristics such as particle size, particle shape, and particle surface coatings that can also contribute to their overall properties. However, every form of nanomaterial depending upon its final core composition has its own unique characteristics, therefore their own way of interactions with bacteria. For example, Jiang et al., while comparing the toxicities of various forms of nanoparticles (Al_2O_3, SiO_2, TiO_2, and ZnO) on several bacterial strains (*B. subtilis*, *E. coli*, and *P. fluorescens*), suggested that all nanoparticles but TiO_2 were highly toxic [20]. ZnO was found to be most toxic with 100 % mortality to all the three strains followed by the Al_2O_3 with killing rate of 57, 36, and 70 % toward *B. subtilis*, *E. coli*, and *P. fluorescens*, respectively, and SiO_2 being the least toxic with mortality: 40 % to *B. subtilis*, 58 % to *E. coli*, and 70 % to *P. fluorescens* [20]. Similarly, in another investigation, among the various forms of nanoparticles (CuO, NiO, ZnO, and Sb_2O_3) that were studied, CuO nanoparticles were found to be most toxic to *E. coli*, *B. subtilis*, and *S. aureus* [21]. In an interesting attempt to utilize engineered nanoparticles as a disinfectant against fish pathogens (*S. iniae* and *E. tarda*), and simultaneously to be able to recollect the nanoparticles back for reuse, researchers developed TiO_2/Fe_3O_4 nanoparticles. TiO_2 with bactericidal properties acted as bactericidal agent and Fe_3O_4 being magnetic, with a magnetization power of 2.9 emu/g aided the separation of TiO_2/Fe_3O_4 from the suspensions using a regular magnet [22]. Similarly, the comparative bactericidal effects of TiO_2 and hybrid Ag-TiO_2 nanoparticles on *B. subtilis* and *P. putida* were evaluated [23]. The authors assessed the toxicity using various parameters such as ratios of Ag and TiO_2, various size distributions, reaction conditions, and doses, using high-throughput bacterial viability assay. The authors concluded that *P. putida* was more resistant than *B. subtilis* due to its lipopolysaccharide mediated high tolerance to nanoparticles. In a recent investigation, Dimkpa et al. reported that CuO nanoparticles were more toxic to soil beneficial rhizosphere isolate *P. chlororaphis* over ZnO nanoparticles [24]. Likewise, Ravi kumar et al., while evaluating the in vitro antibacterial activities of various forms of metal oxide nanoparticles: Al_2O_3, Fe_2O_3, CeO_2, ZrO_2, and MgO against urinary tract pathogens viz., *Pseudomonas* sp., *Enterobacter* sp., *Klebsiella* sp., *E. coli* and *P. morganii* and *S. aureus*, suggested that Al_2O_3 showed the most antibacterial activity against *E. coli*, followed by *Klebsiella* sp. and *P. morganii*, respectively [25]. The authors observed no bactericidal activity against *Pseudomonas* sp. for any of the nanoparticle forms that were used.

Reports also suggest that some forms of nanoparticles can be considered non-toxic or inert, to list a few: Au, Ag_2S, Pt, Pd. Suresh et al., in their separate investigations on the influence of biogenic monodispersed gold nanoparticles and biogenic silver sulfide nanoparticles against Gram-negative (*E. coli* and *S. oneidensis*) and Gram-positive (*B. subtilis*) strains, suggested that these nanoparticles did not elicit significant bactericidal activity and were inert or non toxic [4]. Similarly, in recent investigation, Pelletier et al. reported that even though colloidal CeO_2 (surfactant and template free) nanoparticles showed dose-dependent growth inhibition against bacterial strains *E. coli* and *B. subtilis* and were found to be non-bactericidal [5]. However, similar CeO_2 was neither inhibitory nor bactericidal to the metal reducing bacterium, *S. oneidensis*. Overall, it is evident from the literature that the nanoparticles form and its final constituents can definitely influence nanomaterials behavior thereby its interactions with microbial cells.

1.2.2.2 Influence of Size and Shape Distributions of Nanoparticles

Interactions of engineered nanoparticles with bacteria might also be a result of the particles morphology (size and/or shape), as the greater ratio of surface area to mass occurs as particle size decreases and the properties of nanoparticles are drastically influenced with varying morphologies. Further, smaller particles tend to agglomerate to a greater extent, which may lead to different binding characteristics. For instance, Simon-deckers et al., while evaluating the physico-chemical characteristics of various size and shape distributions of TiO_2 nanoparticles on *C. metallidurans* and *E. coli*, observed a size and or shape influenced effect on the bacteria [26]. The authors observed that as the size decreased, the toxicity is increased and impairment of cell membrane integrity was a major cause of bacterial death. Likewise, Zhang et al., in their studies on the antibacterial effect of zinc oxide nanoparticles on *E. coli*, suggested that the bacterial activity increases with decrease in the particle size and vice versa [27]. The authors additionally observed that the incorporation of dispersants in the form of polyethylene glycol (PEG-400 and PEG-2000) and polyvinylpyrrolidone did not have much influence on the bactericidal activity, however, aided in the enhancement of the overall stability of the nanoparticles suspension. Similarly, Hernandez-Sierra et al., while comparing the antibacterial activity of two different forms and size distributions (silver nanoparticles of 25 nm and zinc nanoparticles of 125 nm) of nanoparticles on *S. mutans*, revealed that silver nanoparticles were found to be more toxic with the minimum inhibitory concentration (MIC) of 4.8 ± 2.7 µg/mL when compared to zinc nanoparticles with a MIC of 500 ± 306 µg/mL [28]. Padmavathy et al. also made similar observations, while assessing the bacterial toxicities of mercaptoethanol capped zinc oxide nanoparticles [29]. The authors believed that the cell death occurred due to the decomposition of cell wall material leading to the leakage of intracellular contents (minerals, proteins, and genetic material) eventually causing cell death.

Shape of the nanoparticle can also have a drastic influence on the overall properties of engineered nanoparticles thereby their interactions with bacteria. Nanoparticles that are irregularly shaped have sharp corners and edges that can be biologically and chemically potent. Atoms at these areas have weaker bonding coordination than their respective bulk counterparts and therefore can bind to foreign molecules with great efficiency. Some of the common shapes of nanoparticles that are currently being used for various purposes are shown in Fig. 1.1. For example, Tam et al., while investigating the antibacterial activity of ZnO nanorods on Gram-negative bacterium, *E. coli*, and Gram-positive bacterium, *B. atrophaeus*, suggested cell membrane damage mediated cell death of the ZnO nanorods and might be due to the release of hydrogen peroxide [30]. Similarly, Pal et al., while evaluating the influence of triangular-shaped silver nanoparticles on *E. coli*, found that triangular silver nanoplates with a {111} lattice plane displayed the most biocidal activity compared to other shapes (spheres and rods) of nanoparticles [31]. In a recent investigation, Pandey et al. reported the remarkable bactericidal potential of very large 500–1000 nm diameter CuO nanoparticles with spiked

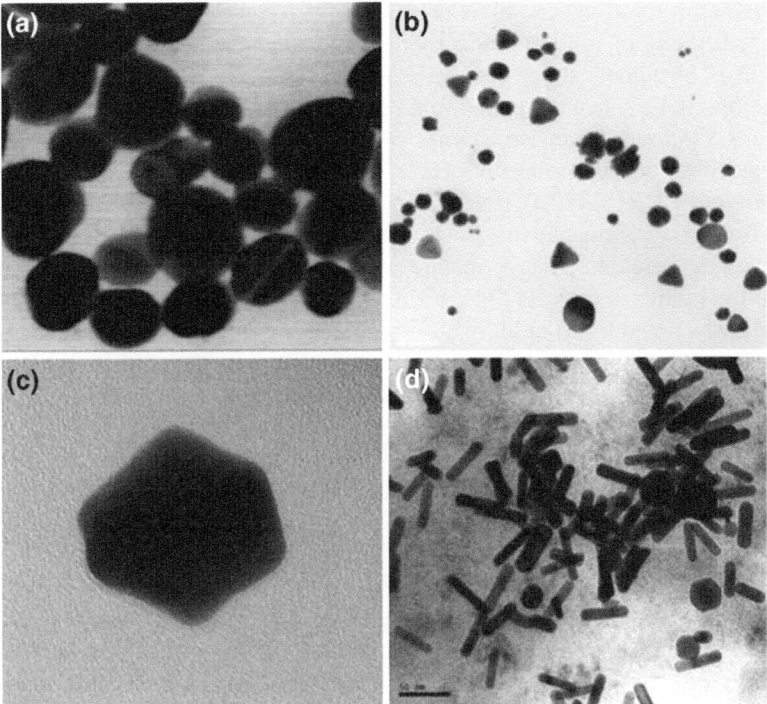

Fig. 1.1 Transmission electron microscopy images of few commonly used shapes of nanoparticles. **a** Spherical silver nanoparticles, **b** triangular gold nanoparticles, **c** hexagonal gold nanoparticles, and **d** gold nanorods

multi-arms against several waterborne disease causing strains: *E. coli, S. typhi, S. aureus,* and *B. subtilis* [32]. The authors showed high bactericidal potential and fast deactivation kinetics killing 10^7 cells within 30 min to 1 h of exposures.

1.2.2.3   Influence of Bacteria Type or Form

Bacterial strain or type was also proven to be an important determinant factor in dictating its own toxicity toward nanomaterials. Differences in the toxicities with respect to wild-type laboratory-cultured bacterial strains versus naturally existing ones and differences in toxicities with regard to Gram-positive versus Gram-negative bacterium have all been observed. Such observed discrepancies could be primarily due to their individual tolerance capacities, sporulation, and differences in their cell membrane structural characteristics. For example, in an investigation on the comparative toxicity impacts of engineered silver nanocrystallites on Gram-negative and Gram-positive bacterium, we noted that bacterial type could also contribute to the observed toxicities. Our results suggested that Gram-negative strains *E. coli* and *S. oneidensis* are more resistant than Gram-positive strain *B.*

subtilis [4]. Similar observations were also made by Ruparelia et al., who noted strain specificity in antibacterial activities while using silver and copper nanoparticles [33]. The authors suggested that *B. subtilis* depicted highest sensitivity to both the nanoparticles when compared to *E. coli* and *S. aureus* and with negligible variation for *S. aureus*. Likewise, Dumas et al., while comparing the toxicity of CdTe quantum dots on Gram-negative (*E. coli* and *P. aeruginosa*) and Gram-positive (*S. aureus* and *B. subtilis*) bacterial strains, revealed that they are significantly more toxic to Gram-negative strains. The authors suggested that even though they observed the release of heavy metal ions (Cd^{2+}), the main cause of toxicity was due to the hydroxyl radicals [34]. Likewise, bacterial species mediated differential toxicity was noted by Li et al., while evaluating the impacts of TiO_2 and Ag-TiO_2 using the bacterium *B. subtilis* and *P. putida*. They observed that the *B. subtilis* was more susceptible when compared to *P. putida* due to the lack of lipopolysaccharide membrane [23].

Reports also exist on the interaction of engineered nanoparticles with environmental microorganisms, the issues that raises serious concern is their toxicity to bacterium that promote plant growth and those that benefit nutrient cycling in soils. Plant growth promoting rhizobacteria such as P. *aeruginosa*, P. *putida*, P. *fluorescens*, B. *subtilis*, and soil nitrogen cycle, nitrifying and denitrifying bacteria have shown varying degrees of inhibition when exposed to engineered nanoparticles in pure culture conditions or aqueous suspensions [35]. Metal oxide nanocrystallites of copper with different size distributions (80–160 nm) were evaluated for antibacterial activity against plant growth promoting *K. pneumoniae*, *P. aeruginosa*, *S. paratyphi*, and *Shigella* strains [36]. Iron- and copper-based nanoparticles are presumed to react with peroxides present in the environment generating free radicals known to be highly toxic to microorganisms such as *P. aeruginosa* [37].

1.3 Mechanisms of Bacterial Toxicity

Depending on the nanoparticles form, physical characteristics, and intrinsic properties, every nanoparticle induces bactericidal activity through one or several of these mechanisms such as membrane damage, dissolution or release of ions, reactive oxygen species (ROS) mediated oxidative damage, and DNA damage. Some of the common proposed mechanisms by which engineered nanoparticles can induce bacterial killing are demonstrated in Fig. 1.2.

1.3.1 Dissolution or Release of Ions

Living organisms do require trace amounts of some heavy metal ions such as iron, cobalt, copper, manganese, molybdenum, and zinc and these elements are called as the essential elements. And some kinds of metallic ions such as the mercury,

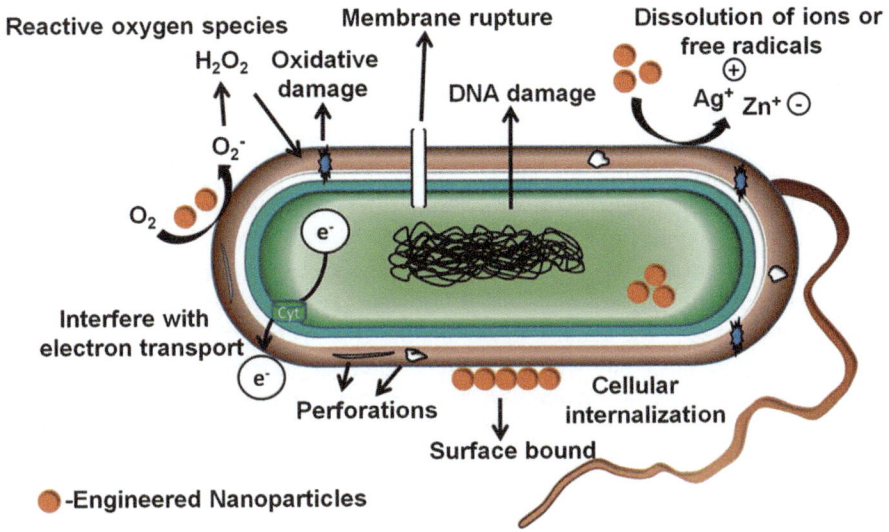

Fig. 1.2 Scheme illustrating some of the proposed mechanisms by which nanoparticles interact with the bacterial cells, along with the modes of toxicity being induced to the bacterial cells. Different nanoparticles might induce toxicity through one or several of these mechanisms. *DNA* deoxyribonucleic acid; *Cty* cytochromes. The image is reproduced from Ref. [50] with permission from The Royal Society of Chemistry

plutonium, cadmium, and lead that are harmful to us are called the non-essential elements. The non-essential elements, primarily, and the essential elements at higher doses are well known to be toxic to majority of the cell organelles. Release of ions and dissolution of ions from the engineered nanoparticles rather than nanoparticles themselves are also considered as a major determinant factor in eliciting nanoparticles toxicity toward bacteria. This phenomenon could be due to the instability of the capping agent initially used and/or degradation or cleavage of the encapping agent upon interaction with biological systems, due to the action of cell protectants in the form of enzymes and/or proteases or other biological materials secreted under such stress conditions.

A limited reports exist on such notion, for example, Lok et al., while investigating the antibacterial activities of small 9 nm diameter spherical silver nanoparticles against silver-resistant bacterium, *E. coli,* suggested that antibacterial activities depend on chemisorbed Ag_+ ions that are formed under extreme oxygen sensitivity conditions. The authors mentioned that the silver nanoparticles aggregated in the media with high electrolyte content, resulting in a loss of antibacterial activities [38]. However, complexation with protein, albumin, could prevent the silver nanoparticles from aggregation, retaining its antibacterial activities. Heavy metal damage due to the release of metal ions was observed when CdSe quantum dot nanoparticles were incubated with the bacterium, *B. subtilis* [39]. The authors further suggested that they have an extrusion mechanism that enables Gram-positive bacteria to pump out cadmium ions [39]. In another investigation, Priester et al., while examining the influence of CdSe quantum dots on *P. aeruginosa,* suggested that the toxicity might be due to release of dissolved heavy metal (Se and Cd) ions [40]. The authors additionally observed impaired growth, membrane damage, cytoplasmic inclusions, and higher levels of ROS [40]. Mahendra et al., while evaluating the influence of different surface engineered CdSe/ZnS core–shell quantum dots (QDs) (QD-carboxyl, QD-polyanionic polymaleic anhydride-alt-1-octadecene, and QD-polycationic polyethylenimine) under acidic (pH ≤ 4) or alkaline (pH ≥ 10) conditions, observed increased bactericidal activity due to the rapid release of cadmium and selenite ions following QD destabilization due to loss of organic coating [41]. The authors stated that the QDs with intact coatings decreased bacterial growth rates but were not bactericidal at near-neutral pH conditions (pH 7) [41]. Similarly, Heinlaan et al. assessed the comparative toxicities of various forms of metal oxide nanocrystallites of ZnO, CuO, and TiO_2 on the bacterium, *V. fischeri* [42]. Based on their observations, TiO_2 nanoparticles were not toxic at all even at higher concentrations (20 g/L), whereas ZnO and CuO nanoparticles were found to be toxic with LC50 values of 1.9 and 79 mg/L, respectively, suggesting that ZnO is more toxic than CuO [42]. The authors revealed that observed toxicity was due to the release of soluble ions which they determined using recombinant Zn and Cu specific sensor bacteria. Similarly, Jiang et al., while comparing the toxicity of various nano- and micro-scaled metal oxide nanoparticles (Al_2O_3, SiO_2, TiO_2, and ZnO) on different bacteria (*E. coli, B. subtilis,* and *P. fluorescens*), suggested that the toxicity completely depends on the differential release of ions for individual nanoparticle, thereby exerting different levels of toxicity [20]. In another investigation, Sotiriou et al. evaluating the interaction of nanosilver upon immobilization on nanostructured silica particles against *E. coli* showed size-dependent release of Ag^+ ions [43]. The authors suggested that the released Ag^+ was dominated when fine particles of dimensions less than 10 nm were used [43]. Therefore, the significance of liberation of dissolved ions should also be carefully examined while considering the bacterial toxicity of engineered nanoparticles.

1.3.2 Reactive Oxygen Species and Free Radicals Mediated Oxidative Damage

ROS are chemically active molecules that contain oxygen, for example, peroxides, and are considered highly reactive due to the presence of unpaired valance electrons. ROS form as a natural by-product of the regular metabolism of oxygen and have important roles including cell signaling, homeostasis, and apoptosis. Generation of ROS and free radicals has been implicated in the toxic response of a number of biological systems to environmental stress, and in the present scenario, the nanoparticles. For example, Adam et al., in their investigation on the comparative potential eco-toxicity of various size distributions and forms: titanium dioxide, silicon dioxide, and zinc oxide on *E. coli* and *B. subtilis*, suggested that the three photosensitive nanomaterials showed varying degree of virulence with bactericidal activity that increased from SiO_2 to TiO_2 to ZnO and was ROS mediated [16]. Additionally, the bacterium, *E. coli*, was found to be more resistant to these effects. Similarly, Ivask et al. showed the varying levels of ROS-mediated ecotoxicities relative to CuO, ZnO, TiO_2, Ag, and fullerene nanomaterials and a set of various recombinant luminescent *E. coli* strains [44]. Applerot et al., in an attempt to understand the influence of particle size on ZnO (microscale to nanoscale) on the bactericidal activity, suggested the reactivity of ZnO with water to be a major cause [45]. The authors suggested that ZnO aqueous suspensions produced higher levels of ROS, namely hydroxyl radicals yielding to oxidative stress-mediated cellular damage, and are size-dependent and microscale particles do not exert any toxicity. Su et al. showed the disruption of bacterial membrane integrity through ROS generation upon treatments with nanohybrids of silver and clay [46]. The authors opined that the plate like clay support with highly concentrated Ag nanoparticles in each silicate unit might disrupt the membrane integrity, increase the production of intracellular ROS, and inactivate the energy-dependent metabolism eventually leading to cell death. Choi et al., while evaluating the influence of four different size distributions (9 ± 5 nm, 15 ± 9 nm, 14 ± 6 nm, 12 ± 4 nm, and 21 ± 14 nm) of silver nanoparticles nitrifying bacteria, observed that the smallest size distributions were the most toxic ones with highest release of ROS [47]. Whereas for the next size distributions of nanoparticles, the ROS correlations were different for different forms rather that size revealing that factor other than ROS does contribute in determining nanosilver toxicity. Recently, contact mediated direct oxidation of the bacterial cell rather than by ROS-mediated oxidative stress was also observed for Nc60 fullerenes [48]. Very recently, Dutta et al. reported the dose-dependent antibacterial action of ZnO nanoparticles against *E. coli* and suggested that it is due to the ROS-induced membrane lipid oxidation [49]. These reports clearly suggest that nanoparticles do induce the liberation of ROS and can enhance ROS-mediated bactericidal activity.

1.4 Summary

An understanding on how engineered nanoparticles interact with microorganisms and how microorganisms may alter the fate, transport, and transformation of engineered nanoparticles in the environment might lead to an increased understanding of the potential environmental impact of commonly used engineered nanoparticles. Ultimately, these studies can guide effective routes for disposal of engineered nanoparticles and potentially guide the development of "ecologically friendly catalysts." Extending this general approach of using well-characterized materials, multiple organisms, and measures of growth and viability to other nanomaterials will be important for understanding the interaction of nanomaterials with living systems and for interpreting the effect and eventual fate of engineered materials in the environment.

References

1. R. Bhattacharya, P. Mukherjee, Adv. Drug Deliv. Rev. **60**, 1289–1306 (2008)
2. A.K. Suresh, M.I. Khan, J. Nanosci. Nanotechnol. **10**(7), 4124–4134 (2005)
3. A.K. Suresh, D.A. Pelletier, W. Wang, J.-W. Moon, B. Gu, N.P. Mortensen, D.P. Allison, D.C. Joy, T.J. Phelps, M.J. Doktycz, Environ. Sci. Technol. **44**, 5210–5215 (2010)
4. D.A. Pelletier, A.K. Suresh, G.A. Holton, C.K. McKeown, W. Wang, B. Gu, N.P. Mortensen, D.P. Allison, D.C. Joy, M.R. Allison, S.D. Brown, T.J. Phelps, M.J. Doktycz, Appl. Environ. Microbiol. **76**, 7981–7989 (2010)
5. A.K. Suresh, Springer Briefs in Molecular Science Biometals, L.L. Barton (ed.) (2012). doi:10.1007/987-94-007-4231-4
6. A.K. Suresh, D.A. Pelletier, M.J. Doktycz, Nanoscale **5**, 463–474 (2013)
7. A. Kahru, H.C. Dubourguier, Toxicology **269**, 105–119 (2010)
8. Y. Ju-Nam, J.R. Lead, Sci. Total Environ. **400**, 396–414 (2008)
9. B.C. Christner, C.E. Morris, C.M. Foreman, R.M. Cai, D.C. Sands, Science **319**, 1214 (2008)
10. E. Rybicki, S. Afr. J. Sci. **86**, 182–186 (1990)
11. T. Gold, Proc. Natl. Acad. Sci. U.S.A. **89**, 6045–6049 (1992)
12. O. Meyer, Zentralblatt für Bakteriologie **281**, 117–119 (1994)
13. D.Y. Lyon, J.D. Fortner, C.M. Sayes, V.L. Colvin, J.B. Hughes, Environ. Toxicol. Chem. **24**, 2757–2762 (2005)
14. G. Jia, H.F. Wang, L. Yan, X. Wang, R.J. Pei, T. Yan, Y.L. Zhao, X.B. Guo, Environ. Sci. Technol. **39**, 1378–1383 (2005)
15. D.Y. Lyon, L.K. Adams, J.C. Falkner, P.J.J. Alvarez, Environ. Sci. Technol. **40**, 4360–4366 (2006)
16. D.Y. Lyon, D.A. Brown, P.J.J. Alvarez, Water Sci. Technol. **57**, 1533–1538 (2008)
17. H. Aoshima, K. Kokubo, S. Shirakawa, M. Ito, S. Yamana, T. Oshima, Biocontrol Sci. **14**, 69–72 (2009)
18. J. Lee, S. Mahendra, P.J.J. Alvarez, ACS Nano **4**, 3580–3590 (2010)
19. S. Kang, M.S. Mauter, M. Elimelech, Environ. Sci. Technol. **43**, 2648–2653 (2009)
20. W. Jiang, H. Mashayekhi, B.S. Xing, Environ. Pollut. **157**, 1619–1625 (2009)
21. Y.W. Baek, Y.J. An, Sci. Total Environ. **409**, 1603–1608 (2011)
22. T.C. Cheng, K.S. Yao, N. Yeh, C.I. Chang, H.C. Hsu, Y.T. Chien, C.Y. Chang, Surf. Coat. Technol. **204**, 1141–1144 (2009)
23. M.H. Li, M.E. Noriega-Trevino, N. Nino-Martinez, C. Marambio-Jones, J.W. Wang, R. Damoiseaux, F. Ruiz, E.M.V. Hoek, Environ. Sci. Technol. **45**, 8989–8995 (2011)
24. C.O. Dimkpa, A. Calder, D.W. Britt, J.E. McLean, A.J. Anderson, Environ. Pollut. **159**, 1749–1756 (2011)
25. S. Ravikumar, R. Gokulakrishnan, P. Boomi, Asian Pac. J Trop. Dis. **2**, 85–89 (2012)
26. A. Simon-Deckers, S. Loo, M. Mayne-L'Hermite, N. Herlin-Boime, N. Menguy, C. Reynaud, B. Gouget, M. Carriere, Environ. Sci. Technol. **43**, 8423–8429 (2009)
27. L.L. Zhang, Y.H. Jiang, Y.L. Ding, M. Povey, D. York, J. Nanopart. Res. **9**, 479–489 (2007)
28. J.F. Hernandez-Sierra, F. Ruiz, D.C.C. Pena, F. Martinez-Gutierrez, A.E. Martinez, A.D.P. Guillen, H. Tapia-Perez, G.A. Martinez-Castanon, Nanomed. Nanotechnol. Biol. Med. **4**, 237–240 (2008)
29. N. Padmavathy, R. Vijayaraghavan, Sci. Technol. Adv. Mater. **9**, 350042 (2008)
30. K.H. Tam, A.B. Djurisic, C.M.N. Chan, Y.Y. Xi, C.W. Tse, Y.H. Leung, W.K. Chan, F.C.C. Leung, D.W.T. Au, Thin Solid Films **516**, 6167–6174 (2008)
31. S. Pal, Y.K. Tak, J.M. Song, Appl. Environ. Microbiol. **73**, 1712–1720 (2007)
32. P. Pandey, S. Merwyn, G.S. Agarwal, B.K. Tripathi, S.C. Pant, J. Nanopart. Res. **14**, 1–13 (2012)
33. J.P. Ruparelia, A.K. Chatteriee, S.P. Duttagupta, S. Mukherji, Acta Biomater. **4**, 707–716 (2008)

34. E. Dumas, C. Gao, D. Suffern, S.E. Bradforth, N.M. Dimitrijevic, J.L. Nadeau, Environ. Sci. Technol. **44**, 1464–1470 (2010)
35. V.K. Mishra, A. Kumar, Digest J. Nanomat. Biostruct. **4**, 587–592 (2009)
36. O. Mahapatra, M. Bhagat, C. Gopalakrishnan, K.D. Arunachalam, J. Exp. Nanosci. **3**, 185–193 (2008)
37. A.M. Saliba, M.C. de Assis, R. Nishi, B. Raymond, E.D. Marques, U.G. Lopes, L. Touqui, M.C. Plotkowski, Microbes Infect. **8**, 450–459 (2006)
38. C.N. Lok, C.M. Ho, R. Chen, Q.Y. He, W.Y. Yu, H. Sun, P.K.H. Tam, J.F. Chiu, C.-M. Che, J. Biol. Inorg. Chem. **12**, 527–534 (2007)
39. S. Silver, Gene **179**, 9–19 (1996)
40. J.H. Priester, P.K. Stoimenov, R.E. Mielke, S.M. Webb, C. Ehrhardt, J.P. Zhang, G.D. Stucky, P.A. Holden, Environ. Sci. Technol. **43**, 2589–2594 (2009)
41. S. Mahendra, H.G. Zhu, V.L. Colvin, P.J. Alvarez, Environ. Sci. Technol. **42**, 9424–9430 (2008)
42. M. Heinlaan, A. Ivask, I. Blinova, H.C. Dubourguier, A. Kahru, Chemosphere **71**, 1308–1316 (2008)
43. G.A. Sotiriou, S.E. Pratsinis, Environ. Sci. Technol. **44**, 5649–5654 (2010)
44. A. Ivask, O. Bondarenko, N. Jepihhina, A. Kahru, Anal. Bioanal. Chem. **398**, 701–716 (2010)
45. G. Applerot, A. Lipovsky, R. Dror, N. Perkas, Y. Nitzan, R. Lubart, A. Gedanken, Adv. Funct. Mater. **19**, 842–852 (2009)
46. H.L. Su, C.C. Chou, D.J. Hung, S.H. Lin, I.C. Pao, J.H. Lin, F.L. Huang, R.X. Dong, J.J. Lin, Biomaterials **30**, 5979–5987 (2009)
47. O. Choi, Z.Q. Hu, Environ. Sci. Technol. **42**, 4583–4588 (2008)
48. D.Y. Lyon, L. Brunet, G.W. Hinkal, M.R. Wiesner, P.J.J. Alvarez, Nano Lett. **8**, 1539–1543 (2008)
49. R.K. Duttaa, B.P. Nenavathua, M.K. Gangishettya, A.V.R. Reddy, Colloids Surf. B **94**, 143–150 (2012)
50. D.M. Aruguete, J.S. Guest, W.W. Yu, N.G. Love, M.F. Hochella, Environ. Chem. **7**, 28–35 (2010)

Chapter 2
Analytical and Physical Characterization Techniques Employed to Assess Microbial Toxicity of Nanoparticles

Abstract Various chemical and/or biological surfactants are often employed while synthesizing nanoparticles, which can drastically contribute to their interactions with bacterial cells thereby toxicity. Analytical and physical characterization techniques that can assess toxicity can potentially guide the proper use of these nanoparticles, for instance, to improve drug formulations for the treatment of infections caused by various multi-drug resistance microorganisms, proper disposal of nanoparticles, etc. Bactericidal activity of a material can be evaluated using several toxicity-assessing techniques. The most commonly used analytical techniques (disk diffusion assay, minimum inhibitory concentration, colony-forming units (CFU) and live–dead staining) and physical characterization techniques [fluorescence spectroscopy, inductively coupled plasma mass spectroscopy, ultra-microtome-based transmission electron microscopy, and atomic force microscopy (AFM)] are described in the current chapter.

Keywords Analytical · Physical · Characterizations · Bactericidal toxicity · Techniques

2.1 Analytical Assays to Determine Microbial Toxicity

The antibacterial activity of any cytotoxic formulation, including nanoparticles, can be assessed using several analytical techniques; some of the commonly used analytical techniques are described below.

2.1.1 Disk Diffusion Assay

A filter paper disk impregnated with a formulation possessing toxic potential when placed on Luria–Bertani agar in a Petri dish, the material based on its diffusion rate tends to diffuse from the disk into the agar. This phenomenon of diffusion or

© The Author(s) 2015
A.K. Suresh, *Co-Relating Metallic Nanoparticle Characteristics and Bacterial Toxicity*, SpringerBriefs in Biometals,
DOI 10.1007/978-3-319-16796-1_2

Fig. 2.1 a Images of an agar plate containing gold nanoparticle formulations impregnated disks showing the diameter of the inhibition zone for *E. coli* (**a**) untreated controls and treated cells (**b**). Diameter of zone of inhibition can be clearly seen

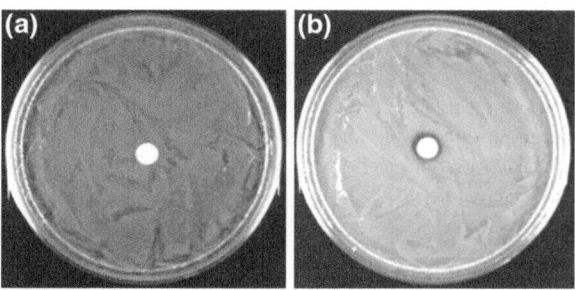

spread of the material in the agar surrounding the disk can also depend on other factors such as solubility, diffusion rate, and molecular size that ultimately determine the area of the materials infiltrating around the disk. When similar experiments are performed with a bacterium placed on the agar, the bacteria will not grow in the region around the disk if it is susceptible to the material. This area of where there is no bacterial growth surrounding the disk is called the diameter of the zone of inhibition, and this assay methodology is called as the disk diffusion assay. The bacterial sensitivity to different nanoparticles could be tested using disk diffusion assay, for which stocks of equal concentrations (25 µg/mL) of the different types of toxic agents (nano constituents, nanoparticles) were first prepared. Then, small disks of uniform size (6 mm diameter) were placed separately in the nanoparticles stock suspensions for five minutes, and the disks were removed carefully using the sterile forceps. The bacterial suspension (100 µL at densities 10^7 cells/mL^{-1}) was spread-plated uniformly on the Luria–Bertani agar Petri dishes using a sterile spreader under sterile conditions, and the nanoparticles impregnated disks are placed in the middle of the plate. The plates were then incubated at 37 °C for 18 h, after which the average diameter of the inhibition zone surrounding the disks was measured using a ruler of 1-mm resolution. As an example, Fig. 2.1 shows the diameter of zone of inhibition performed for *E. coli* using gold nanoparticle–drug formulations. As can be made out from the image, the *E. coli* cells that were untreated, no diameter of zone of inhibition was observed (Fig. 2.1a), whereas for the *E. coli* cells that were treated with nanoparticle formulation, clear zone of inhibition can be seen (Fig. 2.1b) clearly implying that the formulation has bactericidal or killing effect.

As the diameter of the zone of inhibition measurements are performed on an agar Petri dish and the inhibition zone being measured using a ruler, there is a possibility of error; however, the method illustrates the potential analytical assay techniques to assess the toxicity of various toxic agents on different bacterial strains.

2.1.2 Minimum Inhibitory Concentration

Minimum inhibitory concentration is defined as the lowest concentration of a formulation with potent bactericidal activity that can inhibit the visible growth of bacteria after overnight treatment with the formulation. Assessing minimum inhibitory concentration is considered crucial in diagnostic and research purposes to not only confirm the resistance of an organism to the formulation or chemotherapeutics but also to monitor the activity of novel drug formulation. In the standard methodology, to evaluate the minimum inhibitory concentrations of various drug formulations or biologically significant material, the test bacterium was maintained on Luria–Bertani agar Petri dishes, bacterial growth was performed by inoculating a single bacterial colony from the agar plates into fresh liquid Luria–Bertani medium in a culture tube and incubated at 37 °C on a rotary shaker (200 rpm) until the bacterium attains the required growth, which is usually performed by measuring the absorbance at 600 nm. Two hundred microliters of the overnight grown bacterial cells (0.8–1.2 optical density) was inoculated into 10 mL of fresh Luria–Bertani medium. The reaction was performed by transferring 200 µL/well of the above medium into a sterile 100-well bioscreen microtitre or ELISA plate along with the drug formulation, for which the bactericidal activity is desired, at various increasing concentrations separately. The treatments are usually performed in octuplet, and every experiment was repeated at least three times to ensure reproducibility and consistency. The plate was then placed into the bioscreen or ELISA plate reader, and the bacterial growth was monitored every 15 min for 12 h at a wavelength of 600 nm, where the bacteria can be read. Experiments with no drug formulations will be used as control reactions. A greater lag phase and lower maximum absorbance@ 600 nm should be observed for the drug formulation with an increase in the concentration for the bacteria, if the formulation is toxic to that particular bacteria. For example, Fig. 2.2 illustrates the diameter of zone of inhibition performed for the bacterium, E. coli using gold nanoparticle–drug formulations at various increasing concentrations (0.1, 0.2, 0.3, 0.4 and 0.5 µm). As can be seen from the image, the E. coli cells that were untreated had no growth inhibition, where the bacteria are happily growing (Fig. 2.2, bacteria alone line), whereas for the E. coli cells that were treated with various concentrations of nanoparticle formulation, bacterial inhibition was observed (Fig. 2.2).

Fig. 2.2 Bacterial dynamic growth curve for E. coli treated with gold nanoparticle formulation at various increasing concentrations

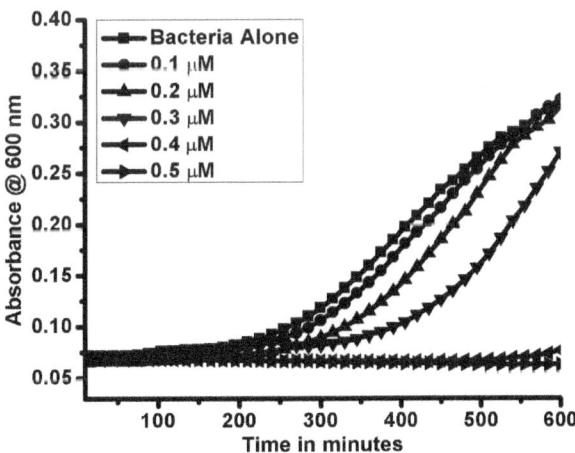

As the concentration of formulation is increased, more amount of particle will available to get adsorbed on to the surface, thereby enhancing the microbial activity. Therefore, a concentration-dependent killing of the bacteria should be observed. However, while performing these bactericidal assays, several other factors from the formulation, such as their size and shape distributions, such as the surface coat, surface charge, and surface properties can contribute and might play a predominant role in dictating potential toxicity by toxic formulate.

2.1.3 Colony-Forming Units

CFU refer to the individual colonies of a microorganism (bacteria, fungi, yeast, or mold). This is used as a measure to assess the number of colonies present in or on the surface of a sample and may be referred as CFU per unit weight, CFU per unit area, or CFU per unit volume depending on the type of sample assessed. To evaluate the number of CFUs, a sample is prepared and spread or poured uniformly on a surface of an agar plate followed by incubation at suitable temperature for few days, until the single colony is grown. To evaluate the effects of any materials on the CFUs, the viability was performed in liquid cultures after treatment with various concentrations of the material. Aliquots will be collected at different time intervals, for example, 0, 2, 8, 12, 18, and 24 h, and serially diluted in the appropriate growth medium, and the dilutions were plated onto Luria–Bertani agar Petri plates. After overnight incubation at 37 °C, the numbers of colonies were counted manually and assessed for the number of colonies grown (CFUs).

2.1.4 Live/Dead Toxicity Assay

The LIVE/DEAD Bacterial Viability assay utilizes mixtures of our SYTO® 9 green-fluorescent nucleic acid stain and the red-fluorescent nucleic acid stain, propidium iodide. These stains differ both in their spectral characteristics and their ability to penetrate healthy bacterial cells. When used alone, the SYTO 9 stain generally labels all bacteria in a population; those with intact membranes and those with damaged membranes. In contrast, propidium iodide penetrates only bacteria with damaged membranes, causing a reduction in the SYTO 9 stain fluorescence when both dyes are present. Thus, with an appropriate mixture of the SYTO 9 and propidium iodide stains, bacteria with intact cell membranes stain fluorescent green, whereas bacteria with damaged membranes stain fluorescent red. The excitation/emission maxima for these dyes are about 480/500 nm for SYTO 9 stain and 490/635 nm for propidium iodide. To determine the LIVE/ DEAD assay for the different nanoparticle samples, bacterial cultures were grown for 3 h in Luria–Bertani medium and subsequently treated with different concentrations of nanoparticle samples for 3–16 h, After which, the microbial suspension

and the stain solution were added to each well of a 96-well micro plate, the plate
was incubated at room temperature in the dark for 15 min, and to quantify the live
and dead cells, the relative fluorescence intensities were measured using a fluorescence plate reader (excitation at 485 nm; emission at 525 and 625 nm). To assess
the percent viability, using the same set wavelength filters, color images can also
be captured using an epi-fluorescence microscope. Green fluorescence represents
the living bacterial cells, and compromised bacterial cells with membrane damage
are stained red. The percent viability of the bacterial cells could then be obtained
from the images using Image J (version 1.4.3) software.

2.2   Physical Characterization Techniques to Ascertain Bacterial Nanoparticle Interactions

The depths of interactions of nanoparticles with bacterial cells can be assessed
using several physical characterization techniques that give away information on
several aspects of nanoparticle bacterial interactions. For instance, nanoparticles
bound to the surface or nanoparticle internalized, probable mode of killing of the
bacteria, biodistribution of nanoparticles within the bacterial cells, mode of damage caused to the bacterial cell (membrane integrity, bacterial membrane rupture,
lesions, and nodules appearing on the bacterial surface, bacterial DNA damage,
etc.). Some of the commonly used physical characterization techniques including
their principle and experimentation details are illustrated below.

2.2.1   Fluorescence Spectroscopy

In fluorescence spectroscopy, the species is first excited, by absorbing a photon,
from its ground electronic state to one of the various vibrational states in the
excited electronic state. Collisions with other molecules cause the excited molecule to lose vibrational energy until it reaches the lowest vibrational state of the
excited electronic state. The molecule then drops down to one of the various vibrational levels of the ground electronic state again, emitting a photon in the process. As molecules may drop down into any of several vibrational levels in the
ground state, the emitted photons will have different energies, and thus frequencies. Therefore, by analyzing the different frequencies of light emitted in fluorescent spectroscopy, along with their relative intensities, the structure of the different
vibrational levels can be determined. In a typical fluorescence (emission) measurement, the excitation wavelength is fixed and the detection wavelength varies, while
in a fluorescence excitation measurement, the detection wavelength is fixed and
the excitation wavelength is varied across a region of interest. An emission map
is measured by recording the emission spectra resulting from a range of excitation
wavelengths and combining them all together. This is a three-dimensional surface

data set: emission intensity as a function of excitation and emission wavelengths and is typically depicted as a contour map.

However, fluorescence spectroscopy can be used an analytical tool only to assess the interaction of selected forms of nanoparticles called the quantum dots that retain the above properties. And is usually based on quenching, refers to any process which decreases the fluorescence intensity of a given substance. A variety of processes can result in quenching, such as excited state reactions, energy transfer, complex formation, and collisional quenching. Quenching is the basis for Forster resonance energy transfer (FRET) assays in biomedical applications, and interaction with a specific molecular biological target is the basis for activatable optical contrast agents for molecular imaging.

2.2.2 Inductively Coupled Plasma Mass Spectroscopy for Quantitative Uptake

Inductively coupled plasma mass spectrometry (ICP-MS) is an analytical technique used for elemental determinations. For the ICP-MS measurements, the sample is usually introduced into the ICP plasma as an aerosol, either by aspirating a liquid or dissolved solid sample into a nebulizer or using a laser to directly convert solid samples into an aerosol. Once the sample aerosol is introduced into the ICP torch, it is completely desolvated and the elements in the aerosol are converted first into gaseous atoms and then ionized toward the end of the plasma. Once the elements in the sample are converted into ions, they are then brought into the mass spectrometer via the interface cones. The interface region in the ICP-MS transmits the ions traveling in the argon sample stream at atmospheric pressure (1–2 torr) into the low-pressure region of the mass spectrometer ($<1 \times 10^{-5}$ torr). The ions from the ICP source are then focused by the electrostatic lenses in the system, and the ions coming from the system are positively charged, so the electrostatic lens, which also has a positive charge, serve to collimate the ion beam and focus it into the entrance aperture or slit of the mass spectrometer. Once the ions enter the mass spectrometer, they are separated by their mass-to-charge ratio. The resolving power (R) of a mass spectrometer is calculated as $R = m/(|m1 - m2|) = m/\Delta m$, where m1 is the mass of one species or isotope and m2 is the mass of the species or isotope it must be separated from; m is the nominal mass.

To determine the nanoparticle concentrations pure nanoparticles suspensions, the sample is directly injected into the ICP plasma, as mentioned above. However, for the bacterial that were treated with nanoparticles, either attached to the surface or internalized, the cells need to be processed. To perform this, the bacterial cells treated with various nanoconstructs are usually exposed for up to 12 h, after which the medium was aspirated, the cells were collected by centrifugation and washed once with PBS, acid digested overnight using 1 mL of 67–70 % HNO_3 until the cell are digested, and analyzed using ICP-MS upon appropriate dilution. Certified standard suspensions (0, 5, 10, 50, 100, 250, 500, 1000, and 2000 ppb) should be run for each experiment as calibrant suspensions.

2.2.3 Dark Field Microscopy

The ability to observe and characterize nanoconstructs and their interaction with biological materials is crucial for nano drug delivery research. Dark field microscopy is an optical microscope technology with optimized focus and alignment of oblique angle illumination (dark field) that can produce a very high signal to noise ratio image. This enables fast observation of the Rayleigh scatter from a wide range of nanoscale materials. For the dark field microscopy assessments, bacterial cells grown overnight, treated with the nanoparticles construct are fixed on a glass slide and mounted with the coverslip. Dark field hyperspectral imaging was performed using a CytoViva dark field microscope system equipped with CytoViva Hyperspectral Imaging System 1.2. Spectral mapping was accomplished using customized ENVI hyperspectral analysis software provided by the manufacturers. First, a library of spectra for particles alone was generated. Each spectra included in the library was sampled from a single pixel imaged with a suitable objective. This library was then mapped onto images of interest by false-coloring a pixel red if it was within 0.1 rad of one of the spectra in the library.

2.2.4 Scanning Electron Microscopy

Scanning electron microscopy (SEM) is a microscope that uses electrons instead of light to form an image is one of the most widely used technique for the characterization of nanostructures. The SEM has many advantages over traditional microscopes, has a large depth of field and higher resolution, which allows larger specimen to be in focus at one time simultaneously closely spaced specimens can be magnified at much higher levels. As SEM uses electromagnets instead of lenses, there is much more control in the degree of magnification. Unlike optical microscopy, this technique not only provides topographical information but also can analyze the chemical composition near the surface. The interaction between the electron beam and the sample gives different types of signals providing detailed information about the surface characteristics, structure, and morphology. When an electron from the beam encounters a nucleus in the sample, the resultant columbic attraction will lead to deflection in the electron's path known as Rutherford elastic scattering. A fraction of these electrons are then backscattered resulting in reemergence from the incident surface. Since the scattering angle depends on the atomic number of the nucleus, the primary electrons arriving at a given detector position produce image yielding topological and compositional data. The high-energy incident electrons can also interact with loosely bound conduction band electrons in the sample. However, the amount of energy given to these secondary electrons as a result of such interactions is small with a very limited range. The secondary electrons produced within a very short distance from the surface escape from the sample giving high-resolution topographical images.

Bacterial treatments with the nanoparticles for the SEM imaging should be performed in a similar manner prepared for TEM imaging and gives information consistent with TEM, the morphological changes of bacterial cells. For SEM analysis, cells (at densities ~10^6 CFU/mL) were treated with desired concentrations of nanoparticle samples for 1–3 h and centrifuged at 3000× g for 30 min. The cell pellets were washed with phosphate-buffered saline (PBS) at least three times and fixed using 2.5 % glutaraldehyde for 30 min. The fixed bacterial cells were washed twice with PBS and treated with 1 % osmium tetroxide for 1 h. After washing with PBS three times, the samples were dehydrated using 30, 50, 70, 80, 90, and 100 % of ethanol treatments, respectively, were dried and gold coated using ion sputter and were imaged under SEM.

2.2.5 *Transmission Electron Microscopy*

Transmission electron microscope (TEM) operates on the basic principles as for the light microscope, however, uses electrons instead of light. And because TEM uses electrons as "light source," the lower wavelength from the electron source makes it possible to get a resolution thousand times better than the light microscope. TEM is often used for high-resolution imaging of thin films of solid samples for structural and compositional analysis. The technique involves (i) irradiation of a thin film by a high-energy electron beam, which is then diffracted by lattices of the crystalline material and propagates in different directions, (ii) imaging and angular distribution analysis of the forward-scattered electrons, and (iii) energy analysis of the emitted X-rays. The topographic information obtained by TEM, in the vicinity of atomic resolution, can be utilized for structural characterization and identification of various forms of nanomaterials, viz. and can discriminate various crystal lattices such as hexagonal, cubic, or lamellar. There are two ways by which one can ascertain nanoparticles interaction with bacteria using TEM; direct imaging and imaging of ultra-thin sections of cell mounted in resin upon ultra-microtome, and are described in detail in the following sections.

2.2.5.1 Direct Imaging Using TEM

The interaction between the bacteria and the nanoconstructs can be illustrated by direct imaging of the bacterial using bright-field TEM imaging of the bacteria treated with the nanoconstructs. Figure 2.3 describes the interaction of silver nanoparticles with the laboratory-standard bacterium *E. coli*. As can be made out from the Fig. 2.3, most of the silver nanoparticles were found attached to the surface of the bacterial cell wall, implying their higher affinity toward the cells. It was obvious that the silver nanoparticles were bound very well throughout the surface of the bacteria and were also able to penetrate the bacteria. Interestingly, it was also observed that the silver nanoparticles looked well separated and spread throughout

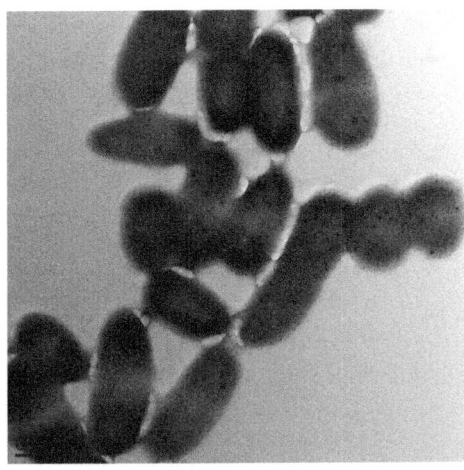

the TEM grid prior to bacterial incubation. Upon incubation with the bacteria, most of the nanoparticles tend to agglomerate and mostly found attached to the surface the bacteria. As demonstrated by electron microscopy, interaction with silver nanoparticles resulted in perforations in the cell wall, contributing to the enhanced antibacterial effects of the nanoparticles. Additionally, clusters of particles are seen throughout the bacterial surface. However, using direct imaging technique, the biodistribution or cell distribution of the particles within the cells cannot be assessed, therefore trans-sectional TEM, where the nanoconstructs treated cells are mounted into resin, ultra-thin sections of which is made using platinum blade and then imaged under TEM. Details of which are illustrated in the later section.

2.2.5.2   Ultra-Microtome-Based Trans-Sectional Imaging Using TEM

Trans-sectional-based transmission electron microscopy is also becoming an emerging tool to assess the local biodistribution of the various nanoparticle constructs within cells or biological tissues. After uploading of the cells or tissue with the desired nanovectors, the cells or tissue are sectioned using a very fine blade (diamond) that can cut or make very thin sections of the cells or tissues up to nanometer scale lengths. For these experiments, the bacterial cells treated with nanoparticle constructs were collected by centrifugation, followed by washing with PBS, fixed with 2 % glutaraldehyde in 0.1 M cacodylate buffer ($Na(CH_3)\cdot 2AsO_2\cdot 3H_2O$), pH 7.2, at 4 °C overnight. The following day, the bacterial cells were washed three times with 0.1 M cacodylate buffer, post-fixed with 1 % OsO_4 in 0.1 M cacodylate buffer for 30 min and washed three times with 0.1 M cacodylate buffer. The samples were then dehydrated using 60, 70, 80 and 95 % ethanol and 100 % absolute ethanol (twice), propylene oxide (twice) and were leveled in propylene oxide/eponate (1:1) overnight at room temperature under a sealed environment. The following day, the vials were level open until the propylene oxide evaporated (2–3 h). The samples were infiltrated with 100 % eponate and polymerized at 64 °C for 48 h. Ultra-thin Sections (70 nm thick) were cut using a Leica Ultra-cut UCT ultra-microtome equipped with a diamond knife, and the sections were picked up onto 200 mesh copper TEM grids. The grids were stained with 2 % uranyl acetate for 10 min followed by Reynold's lead citrate staining for a minute and were imaged using a TEM.

2.2.6 Atomic Force Microscopy

Atomic force microscope (AFM) is a form of scanning probe microscopy (SPM) wherein a small probe is scanned across the sample to obtain information about the sample's surface. The information gathered by such interaction can be as simple as physical topography or as diverse as measurements of the material's physical, magnetic, or chemical properties. AFM has the advantage of imaging almost any type of surface such as thick film coatings, ceramics, composites, glasses, synthetic and biological membranes, microorganisms and cells, biomaterials, metals, and polymers. AFM is being applied to study various phenomena such as the abrasion, adhesion, cleaning, corrosion, etching, friction, lubrication, plating, and polishing. The AFM probe has a very sharp tip, often less than 100 Å diameter, at the end of a small cantilever beam. The probe is attached to a piezoelectric scanner tube. Interatomic forces between the probe tip and the sample surface cause the cantilever to deflect as the sample's surface topography or other properties change. A laser light reflected from the back of the cantilever measures the deflection of the cantilever. This information is fed back to a computer, which generates a map of topography and/or other properties of interest. However, based on the type of application, different operation modes of AFM are used like the contact mode, where the AFM measures the hard-sphere repulsion forces between the tip and the sample; non-contact mode, where the AFM derives topographic images from measurements of attractive forces; the tip does not touch the sample; and the tapping mode, where the cantilever is driven to oscillate up and down at near its resonance frequency by a small piezoelectric element mounted in the AFM tip holder similar to non-contact mode. To be able to perform AFM, the bacterial cells should be mounted or immobilized onto a solid mica substrate coated with gelatin, so as to aid imaging, details of which is described below.

2.2.6.1 Immobilization of Bacteria onto Gelatin-Coated Mica

Immobilization is a commonly used technique for the physical or chemical fixation of materials (e.g., cells, organelles, proteins, molecules) onto a solid substrate, into a solid matrix, or retained by a membrane, in order to increase their stability and make possible their repeated or continued use. Here, we immobilize bacterial cells on to gelatin-coated mica so as to detect them using the AFM, for which

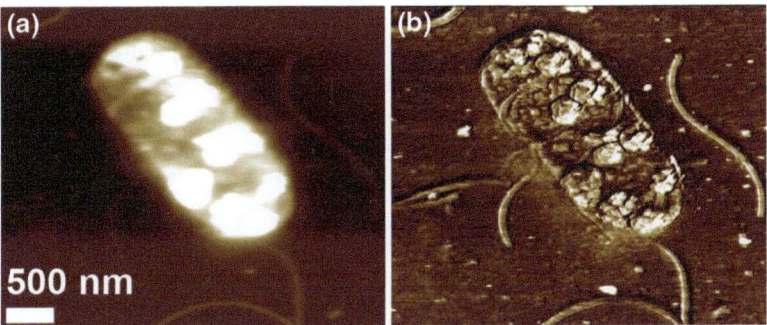

Fig. 2.4 AFM images showing the interaction of the bacterium, *E. coli*. **a** Topographical. **b** Deflective images with silver nanoparticles

freshly cleaved mica surfaces were dipped into 0.5 % gelatin prepared in Milli Q water at 60 °C and dried overnight. A total of 100 μL of the sample analyte was applied to the gelatin-coated mica surface, allowed to stand for 10 min, rinsed in Milli Q water, and placed in the cell for AFM imaging.

AFM measurements were performed to clearly understand the depth of interaction between the bacteria and the silver nanoparticles. Several reports exist on the suitability of AFM for investigating cell structure and morphology of both human cells as well as bacterial cells. AFM is an appropriate technique for elucidating the action of bactericides on bacterial cells. In fact, several investigations performed by various groups suggested noticeable significant changes in the cell membrane morphology upon treatment with different bactericidal agents and could be monitored with ease using AFM. Bacteria with no nanoparticles should be used as controls where the cells looked healthy, intact with no perforations. Figure 2.4 shows the interaction of laboratory-standard *E. coli* with silver nanoparticles; as assessed from the figure, *E. coli* cells were significantly damaged with several perforations and pitches throughout the surface upon treatment. The Ag nanoparticles made long scars on the bacterial surface completely tearing the membrane showing its high killing efficiency. Despite the fact that the mechanism of this interaction is still unanswered, the nanoparticles might cause perforations thereby structural changes, degradations, and finally cell death.

2.3 Summary

The most commonly used analytical (disk diffusion assay, minimum inhibitory concentration, CFU and live–dead staining) and physical characterization (fluorescence spectroscopy, inductively coupled plasma mass spectroscopy, ultra-microtome-based transmission electron microscopy, and AFM) techniques that will shed information on the probable modes of interaction of nanoparticles with bacterial cells are discussed. The detailed methodologies of each and every technique including sample preparation and processing illustrating suitable examples are presented. A better understanding on the interaction of nanoparticles with bacterial cells are considered crucial for the intended proper use of nanoparticles either as better bactericidal agents and or for biomedical purposes.

References

1. A.K. Suresh, D.A. Pelletier, M.J. Doktycz, Nanoscale **5**, 463–474 (2013)
2. A.K. Suresh, M.I. Khan MI, J. Nanosci. Nanotechnol. **10**(7), 4124–4134 (2005)
3. A.K. Suresh, D.A. Pelletier, W. Wang, J.-W. Moon, B. Gu, N.P. Mortensen, D.P. Allison, D.C. Joy, T.J. Phelps, M.J. Doktycz, Environ. Sci. Technol. **44**, 5210–5215 (2010)
4. D.A. Pelletier, A.K. Suresh, G.A. Holton, C.K. McKeown, W. Wang, B. Gu, N.P. Mortensen, D.P. Allison, D.C. Joy, M.R. Allison, S.D. Brown, T.J. Phelps, M.J. Doktycz, Appl. Environ. Microbiol. **76**, 7981–7989 (2010)
5. A.K. Suresh, Springer Briefs in Molecular Science Biometals, L.L. Barton (ed.) (2012). doi:10.1007/987-94-007-4231-4

Chapter 3
Toxicity of Metal Oxide Nanoparticles: Bactericidal Activity and Stress Response

Abstract Nanoparticles are being developed for several researches as well as for medicinal and engineering implications. Following them, a host of new potential health issues due to their size dependent larger surface area and high reactivity. In this chapter, an introduction on the likely interactions of nanoparticles with biotic environment, various possibilities of these man-made nanoparticles coming in contact with the environment, and thereby consequences will be discussed. This will be illustrated using our recent work on the effects of various sizes of engineered zinc oxide nanoparticles on the growth and viability of *Escherichia coli* and *Bacillus subtilis*. The relation between the growth inhibition and reactive oxygen species (ROS) generation and up- and/or down-regulation of transcriptional stress genome using *E. coli* and *B. subtilis* will be discussed. How zinc oxide nanoparticles were synthesized by solvent-free hydrothermal-based approach so as to eliminate cross-contaminants from the use of toxic solvents and surfactants. Further, utilization of advanced technique such as the transmission electron microscopy (TEM) and microarray-based transcriptional profiling to evaluate the bacterial response mechanisms will be described.

Keywords Bactericidal · *E. coli* · B. subtilis · Zinc oxide · Nanorods · Stress response

3.1 Metal Oxide Nanoparticles and Bacteria

The intrinsic size- and shape-dependent physicochemical, optoelectronic, magnetic, catalytic, and biological properties of nanomaterials have been found to vary significantly as compared to that of bulk materials [1]. These unique properties can in turn strongly modulate changes in the color, thermal behavior, material strength, conductivity, solubility, and catalytic activity that selectively contribute to numerous implications such as in effective heterogeneous catalysis, novel probes for sensing and cell imaging, and drug delivery applications. Though the benefit of

© The Author(s) 2015   27
A.K. Suresh, *Co-Relating Metallic Nanoparticle Characteristics
and Bacterial Toxicity*, SpringerBriefs in Biometals,
DOI 10.1007/978-3-319-16796-1_3

engineered nanomaterials has long been known, it is until recently that they are being synthesized in huge quantities and finding applications in a wide range of commercial products, including cosmetics, medicines, clothing, electronics, paint and fuel additives, and engineering [1]. For example, nanoparticles associated with polymers, metal or metal oxides, liposomes, micelles, dendrimers, or metal sulfides are being considered for combating diseases such as cancer [2] or fighting bacterial pathogens [1, 3].

Zinc oxide (ZnO) nanoparticles are a prime example of a metal oxide nanomaterial that is being developed for a wide range of novel industrial and biomedical uses due to its large surface area and semiconductor nature that in turn lead to several interesting properties. It is used extensively as an abrasive semiconductor for manufacturing, as a catalyst for automobile exhaust, and as a photocatalyst because of its piezoelectric behavior [4] in pigments due to its ceramic properties [5]. The conductivity of ZnO nanoparticles changes as molecules get absorbed onto their surface, and this specific property allows them to be used in solid-state gas sensors [6]. Due to its antibacterial activity, it has long been used in several dermatological applications in the form of lotions, creams, and ointments [7]. Apart from the above-mentioned applications, ZnO nanoparticles have also been implemented in various other applications such as solar cells, transparent electrodes, electroluminescent devices, and ultraviolet laser diodes.

The proliferation of nanotechnology has prompted researchers over the safety of these engineered nanomaterials to both mankind as well as the environment. The rate at which research in nanoscience is progressing and being utilized for some or other applications, it is almost inevitable that living beings will be exposed to these nanomaterials [8]. The same properties that make these nanoparticles useful in various applications can potentially have adverse effects on the environment. The potential toxicity of nanomaterials has been recognized and known to elicit either inflammatory response, oxidative stress, lipid peroxidation, mitochondrial dysfunction, or DNA damage [1]. Nevertheless, a thorough understanding of the hazards associated with individual nanomaterial type might reduce damage caused to the biosphere or toxic effects on health [9, 10].

Majority of the reports relating to toxicity assessments of metal oxide nanoparticles have focused on eukaryotic cells, with emphasis on cancer cell lines. Their influence on microbial systems has not been recognized except fewer recent studies that have focused on the effects of ZnO and other metal oxide nanoparticles on bacterial systems [1]. However, these reports lack a systematic strategy or a standardized route to show the microbial interactions with engineered nanomaterials and the exact mechanism of the bacterial response to these nanomaterials. Therefore, the present chapter illustrates the synthesis and characterization of zinc oxide nanoparticles, and the different approaches appropriate for assessing bacterial toxicity using well-characterized materials and standard bacterial assay systems. Specifically, examining how the presence of ZnO nanoparticles affects growth and viability of *Escherichia coli* and *Bacillus subtilis* by performing bacterial growth with and without the presence of nanoparticles. Additionally, the use of studies such as the reactive oxygen species (ROS) generation, transmission electron microscopy (TEM) imaging, and microarray-based transcriptional profiling that potentially reveal in detailed mechanism of their interaction and the genetic response of the bacteria to zinc nanoparticle stress is ascribed.

3.2 Synthesis and Characterization of Zinc Oxide Nanoparticles

ZnO nanoparticles were synthesized by forced hydrolysis of zinc acetate dehydrate.

$$Zn(CH_3COO)_2 + H_2O \rightarrow HO\text{--}Zn\text{--}(CH_3COO) + CH_3COOH$$
$$HO\text{--}Zn\text{--}(CH_3COO) + H_2O \rightarrow Zn(OH)_2 + CH_3COOH$$
$$Zn(OH)_2 \rightarrow ZnO + H_2O$$

Briefly, 5.5 g of $Zn(CH_3COO)_2 \cdot 2H_2O$ and 1 mL of H_2O were dissolved into 100 mL diethylene glycol (DEG) at 60 °C to form an yellow suspension, which was then rapidly added into another 150 mL of boiling DEG. The conversion of opaque yellow-orange color to white-yellowish suspension indicates the formation of ZnO nanoparticles. The reaction was kept under reflux for 15 min and then cooled down to room temperature. The synthesized ZnO nanoparticles were collected by centrifugation and purified by washing couple of times with Milli Q and dialysis against Milli Q to remove other soluble impurities.

The above-obtained pure zinc oxide nanoparticles were characterized in terms of its purity, crystallinity, morphology, dimensions, and surface charge using various physical characterization techniques. For instance, UV–vis absorbance measurements were recorded on a CARY 100 Bio Spectrophotometer (Varian Instruments, CA) operated at a resolution of 1 nm and revealed intense peaks at

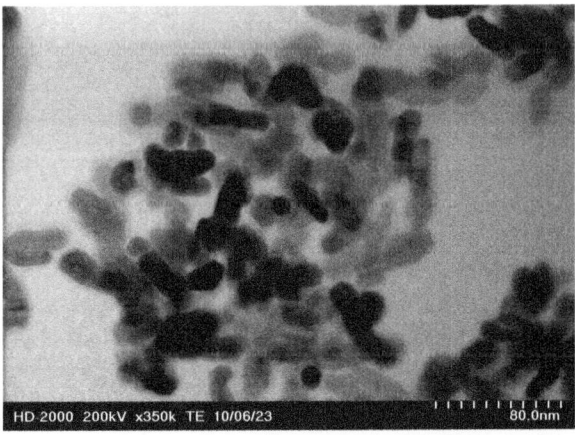

Fig. 3.1 Transmission electron microscopy image of ZnO nanoparticles

355 nm indicating the presence of zinc oxide nanoparticles. The size and shape distribution of the ZnO nanoparticles were assessed by TEM micrograph imaging for the samples drop-coated on carbon-coated copper grids and were taken on a Hitachi HD-2000 STEM operated at an acceleration voltage of 200 kV. TEM imaging revealed them to be predominantly rod shaped, with a long distance of $\sim 22 \pm 3$ nm and a short distance of $\sim 9 \pm 2$ nm (Fig. 3.1). The TEM images showed crystalline particles.

3.3 Bactericidal Assessment of Zinc Oxide Nanoparticles

Details on the description of the various toxicity evaluation methods along with the principle mechanism are described earlier. Therefore, only the bactericidal toxicity assessments performed using the zinc oxide nanoparticles are described below.

3.3.1 Disk Diffusion Assay

When a filter paper disk impregnated with a material possessing toxic properties is placed on Luria–Bertani agar, the materials based on its diffusibility tend to diffuse from the disk into the agar. This diffusion will place the material in the agar surrounding the disk, which further depends on factors such as solubility, diffusion rate, and molecular size that ultimately determine the distance of the area of material infiltration around the disk. If a bacterium is placed on the agar, it will not grow in the region around the disk if it is susceptible to the material. This area of no bacterial growth surrounding the disk is known as the "*diameter of the zone of inhibition (DIZ)*," and the method is called the disk diffusion test. The bacterial sensitivity to different engineered silver nanoparticles was tested using disk diffusion test, for which small disks of uniform size (6 mm diameter) were placed separately in the zinc oxide nanoparticles suspensions for 5 min and the disks were removed carefully using the sterile forceps. The bactericidal suspension (100 μL of 10^4–10^5 CFU mL^{-1}) was spread and plated uniformly on the Luria–Bertani agar Petri dishes using a sterile spreader under sterile conditions, before placing the disks on the plate. The plates were then incubated at 37 °C for 18 h, after which the average diameter of the inhibition zone surrounding the disks was measured using a ruler of 1 mm resolution.

When the antibacterial activity for zinc oxide nanoparticles on the *B. subtilis* bacteria was compared using the DIZ in disk diffusion assay, the strains susceptible to disinfectants showed larger DIZ (see Fig. 3.2), whereas if the cells are resistant, they exhibit smaller or no DIZ. The disks with zinc oxide nanoparticles were surrounded by a larger DIZ for *B. subtilis* bacterial strain.

Fig. 3.2   Image of an agar
plate containing zinc oxide
nanoparticle (50 mg/mL)
impregnated disks showing
the diameter of the inhibition
zone for *B. subtilis*

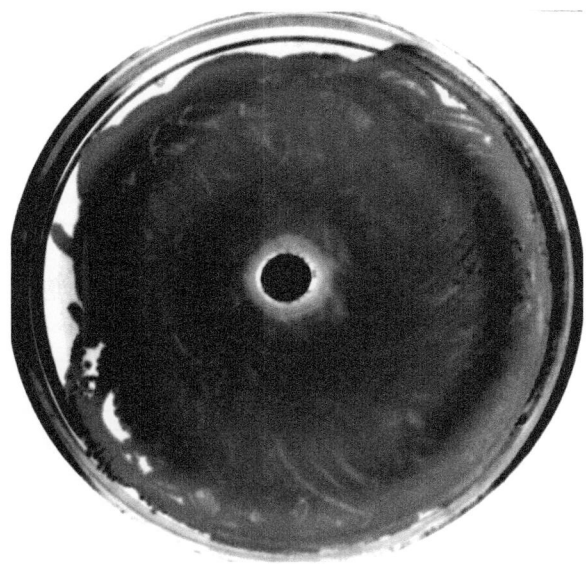

3.3.2 Determination of Minimum Inhibitory Concentration

Minimum inhibitory concentration (MIC), defined as the lowest concentration of
a compound that inhibits the growth of an organism [11], was determined for *B.
subtilis* using 100-well bioscreen plates containing 200 µL of the bacterial cultures (~0.098 OD) and varying concentrations of ZnO nanoparticles or zinc chloride (positive controls). Bacterial growth was recorded every 15 min for 8 h at a
wavelength of 600 nm in a bioscreen plate reader (Thermo Labsystems, Finland).
Each treatment was performed in octuplets, and every experiment was repeated at
least three times to ensure reproducibility. Bacterial growth and viability measurements upon exposure to ZnO nanoparticles were performed. To assess the antibacterial activity, the MIC for *B. subtilis* was determined with varying concentrations
of the ZnO nanoparticles (Fig. 3.3) in a 100-well bioscreen plate, in bioscreen
plate reader and the bacterial growth was monitored every 15 min for 8 h at a
wavelength of 600 nm. A greater lag phase was observed with increasing concentrations of the nanoparticles, with a MIC of 50 mg/L.

3.3.3 Live/Dead Viability Assay

B. subtilis cultures grown to logarithmic phase in Luria–Bertani medium, treated
with varying concentrations of ZnO nanoparticles. Following exposure, the impact
on bacterial membrane integrity was assessed using Live/Dead BacLight Bacterial

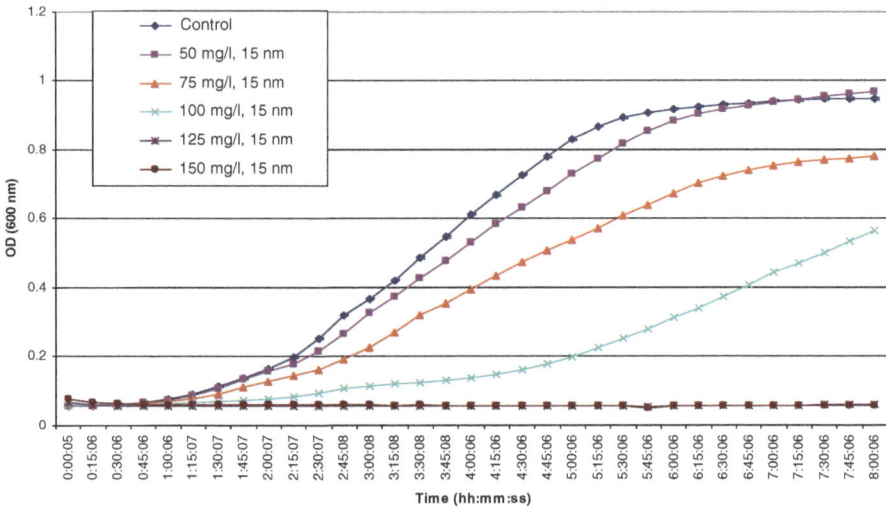

Fig. 3.3 Bacterial growth curves of *B. subtilis* treated with various concentrations of ZnO nanoparticles

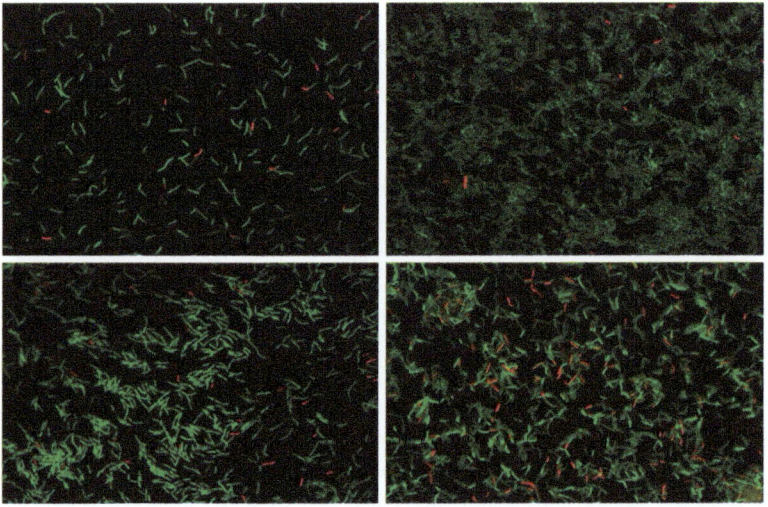

Fig. 3.4 Live/dead staining assay for *B. subtilis* treated with zinc oxide nanoparticles. *B. subtilis* cells showing differential staining of cells with green SYTO9 (stains all cells, live or dead) and red propidium iodide (stains dead or membrane compromised cells). *Top row* untreated control cells at 0 (*left*) and 2 h (*right*), whereas *bottom row* is for cells treated with ZnO nanoparticles at similar times

Viability Kit (Invitrogen) following manufacturers instructions. To quantify the relative number of live and dead cells, the relative fluorescence intensities were measured using a fluorescence plate reader at an excitation of 485 nm; the emission was recorded at 525 and 625 nm. To further understand the potential bactericidal nature of the ZnO nanoparticles, log phase bacteria, grown in Luria Bertani medium, were treated with varying concentrations of ZnO nanoparticles for 15 h and analyzed using a Live/Dead BacLight Bacterial Viability Kit. The observed fractions of "dead" cells were similar to the trends observed in the MIC assays discussed above (Fig. 3.4).

Overall, the physical characterizations of the different zinc oxide nanoparticles indicate the heterogeneities and chemical properties associated with the samples that may need to be accounted for when interpreting toxicity data. Clearly, even with refined synthesis procedures, a range of physical structures is present. These structures may have different biological reactivity that can complicate interpretations. Further, the point of zero charge of the particles is near neutral pH values and can cause the particles to agglomerate at pH values optimal for bacterial growth. This point of zero charge can also shift in different media used for bacterial growth indicating the presence of other media components competing for binding sites on the particles.

3.4   Mechanism of Toxicity

To evaluate the molecular mechanisms that underlie bacterial response to the zinc oxide nanoparticles, a series of imaging experiments and molecular genomic analyses were performed. These assessments focused on the effects of the 15 ± 4.3 nm sized zinc oxide nanoparticles on *B. subtilis* due to its demonstrated inhibition on cell growth. Although such analyses are often involved, an analysis of the molecular mechanisms can ultimately be used to classify bacterial response mechanisms.

3.4.1   Mode of Interaction Based on Transmission Electron Microscopy

To assess whether there is a potential mode of interaction between the zinc oxide nanoparticles and *B. subtilis*, TEM imaging experiments of the nanoparticle-treated bacteria were performed. Effective imaging required refinement of the experimental protocol. Initial experiments involved placing a droplet of bacteria/nanoparticles on the carbon-coated grids and air-drying. However, this resulted in either a film or precipitate from the media obscuring nanoparticle interactions with the bacteria. To eliminate this problem, the nanoparticle/bacteria culture was pelleted and resuspended in water prior to placing on the grid. However, images of these samples showed that nanoparticles in the solution migrate toward the edges of the bacteria forming a halo around the perimeter of the bacteria, presumably during drying of the sample. By allowing a droplet of the bacteria containing solution to first settle on the copper or nickel-/carbon-coated grid and then rinsing the surface by plunging once into water eliminated this artifact and removed unbound nanoparticles and unbound bacteria.

Transmission electron microscopy (TEM) imaging of the nanoparticles for the nanoparticle-treated bacteria was carried out by placing a 5 µL droplet on the grid,

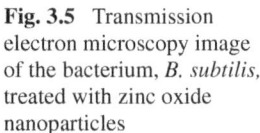

Fig. 3.5 Transmission electron microscopy image of the bacterium, *B. subtilis,* treated with zinc oxide nanoparticles

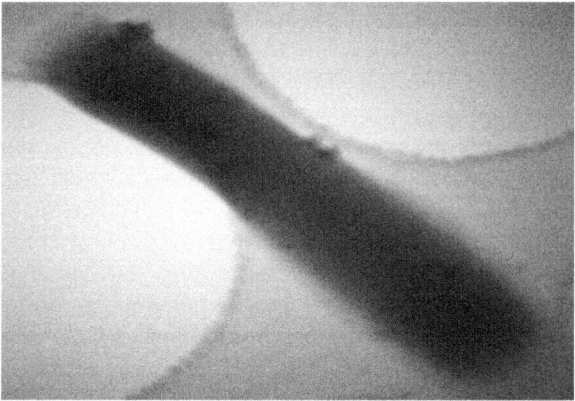

incubating for 8 min, rinsing by plunging with Milli Q water, and air-drying. The mode of potential interaction between the zinc oxide nanoparticles and the bacterium, *B. subtilis,* was determined by STEM micrograph imaging. As can be made out from Fig. 3.5, a representative image of *B. subtilis* treated with ZnO nanoparticles, clearly suggesting that the nanoparticles were adsorbed to, and also might have penetrated, the bacterial cells.

3.4.2 Monitoring Reactive Oxygen Species (ROS) Production

Reactive oxygen species (ROS) are chemically reactive molecules such as peroxides that contain oxygen. ROS are highly reactive due to the presence of unpaired valence shell electrons. ROS form as a natural by-product of the normal metabolism of oxygen and have important roles in cell signaling, homeostasis, and also apoptosis. However, during times of environmental stress, in the present case, in the form of nanoparticles, ROS levels are known to increase drastically which might result in significant damage to cell structures. This cumulates into a situation known as oxidative stress. ROS production can be monitored using various analytical techniques. In our study, the ROS production upon exposure of the bacterial suspension to smallest zinc oxide nanoparticles was monitored by change in the color of 2,3-bis(2-methoxy-4-nitro-5-sulfophenyl)-2H-tetrazolium-5-carboxanilide (XTT) due to the reduction of superoxide (O^{2-}) to XTT-formazan. The ROS production upon exposure of the bacterial suspension to ZnO nanoparticles was monitored by change in the color of 2,3-bis(2-methoxy-4-nitro-5-sulfophenyl)-2H-tetrazolium-5-carboxanilide (XTT) due to the reduction of superoxide (O^{2-}) to XTT-formazan.

Briefly, in a 96-well plate, samples containing various concentrations of the ZnO nanoparticles, 100 µM XTT in 200 µL of appropriate medium, were monitored for change in the absorbance at 470 nm using a spectrophotometer at various time intervals, which is indicative of superoxide production. The generation of ROS upon bacterial interaction with ZnO nanoparticles was examined using an XTT assay, which yields a colorimetric signal when reduced by superoxides. Using this assay, involving *E. coli,* we found a clear signal with ZnO that gradually increased upon increasing nanoparticle concentration suggesting that ZnO nanoparticles does, in fact, cause oxidative stress by generating superoxide. ROS is best known to implicate toxicity to several prokaryotic and eukaryotic systems upon interaction with metal/metal oxide nanoparticles. ROS, either in the form of superoxide radical (O^{2-}), hydrogen peroxide (H_2O_2), or hydroxyl radical (OḢ) causes oxidative stress, thereby damages DNA, cell membranes, cellular proteins, and finally leading to cell death. The presence of ROS was examined using an XTT assay, which yields a colorimetric signal when reduced by superoxides.

3.4.3 Transcriptional Microarray Analysis

Microarray is a multiplex technology used in molecular biology and in medicine. It consists of an arrayed series of thousands of microscopic spots of DNA oligonucleotides, called features, each containing picomoles of a specific DNA sequence. This can be a short section of a gene or other DNA element that is used as probes to hybridize a cDNA or cRNA sample (called target) under high-stringency conditions. Probe-target hybridization is usually detected and quantified by detection of either fluorophore- or silver- or chemiluminescence-labeled targets to determine relative abundance of nucleic acid sequences in the target. In standard microarrays, the probes are attached to solid surface by a covalent linking to a chemical matrix using cross-linkers such as epoxy-silane, aminosilane, lysine, and polyacrylamide. The solid surface can be a glass or a silicon chip, commonly known as gene chips. Microarrays can be used to measure changes in expression levels and to detect single nucleotide polymorphisms, in genotyping or in resequencing mutant genomes. DNA microarrays can be used to detect DNA (as in comparative genomic hybridization) or to detect RNA (most commonly as cDNA after reverse transcription) that may or may not be translated into proteins. The process of measuring gene expression via cDNA is called expression analysis or expression profiling. For discovery of genetic-based response mechanisms, the global transcriptomic response of *E. coli* upon exposure to zinc oxide nanoparticles was assessed using whole-genome microarray analysis and compared to treatments with zinc chloride or Milli Q water, as described by us earlier for cerium oxide nanoparticles [1].

3.4.3.1 Microarray Hybridization Methodology

For microarray experiments, an overnight grown culture of *E. coli* was used to inoculate into 100 mL of pre-warmed Luria–Bertani medium to an optical density of ~0.096 (at 600 nm) and incubated at 37 °C on a shaker at 200 rpm until mid-log phase (at 600 nm ~ 0.5). Cultures were treated separately with either pre-warmed zinc oxide nanoparticles or zinc chloride suspension at a little higher concentration than required to induce minimum inhibition, ~100 mg/L and Milli Q water alone. After one-hour treatment, cells were collected by centrifugation (5000 g, 2 min at 4 °C) and snap-freezing using liquid nitrogen. Three separate controls and three experimental cultures were determined for every condition. Incubating the cells with 1 mg/mL of lysozyme to lyse the cells isolated total cellular RNA. Purified, fluorescently labeled cDNA was hybridized to *E. coli* K12 gene expression 4 × 72 K arrays using a Nimblegen Hybridization System, according to the manufacturer's instructions. Microarrays were washed using buffers of increasing stringency, microarray mixes were removed in 42 °C Nimbelgen Wash Buffer I, washed manually in room temperature buffers; Wash Buffer I for 2 min, Wash Buffer II for 1 min, Wash Buffer III for 15 s, dried for 80 s using a Maui Wash System, scanned with a Surescan high-resolution DNA microarray scanner and the images were quantified using the Nimblescan software. Raw microarray data were log_2 transformed and imported into the statistical analysis software JMP Genomics 4.0. Microarray data were normalized using the Lowess normalization algorithm within JMP Genomics, and an analysis of variance (ANOVA) was performed to determine significant differences in gene expression levels between conditions and time points using the FDR testing method ($p < 0.01$).

3.4.3.2 Stress Genomic Analysis

For the identification of genomics-based toxicity response mechanism, the global transcriptomics of *E. coli*-treated 15 ± 4.3 nm sized zinc oxide nanoparticles was evaluated based on whole genomic microarray analysis and compared to similar treatments with zinc chloride or water. In the microarray experiments, there was only a slight impact on cell growth and no appreciable differences in culture responses to the respective treatments. To understand the mechanism explaining the toxicity of the nanoparticles to the microbes, we conducted a microarray experiment to gain a system-level view of the genetic response of the microbes toward ZnO exposure. Preliminary data showed a number of genes that were differentially expressed as soon as 5 min after exposure to ZnO and $ZnCl_2$. Smaller sets of genes were regulated at 1 h. The comparison of gene expression levels at 5 min and 1 h after shock with ZnO and $ZnCl_2$ showed differential gene regulation suggesting the unique toxicity of ZnO nanoparticles toward *E. coli* (Fig. 3.6). Pair wise co-relations of the gene expression levels of the bacterium, *E. coli*, at 5 min and 1 h after shock treatments with ZnO nanoparticles and $ZnCl_2$ are given in Fig. 3.6. Genomic and proteomic tools provide a unique opportunity to not just develop markers for toxicological studies [12, 13] but also probe for alternative mechanisms responsible in toxicity as we have done with our study. Our system-level gene expression experiment for *E. coli* is a unique approach in elucidating the other aspects involved with nanoparticle toxicity. The microarray data have provided a wealth of information that will help put together the bigger picture of

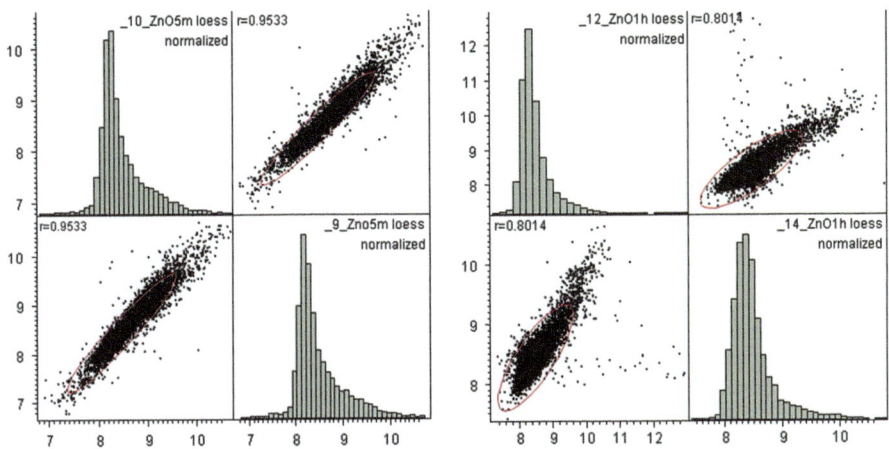

Fig. 3.6 Pair wise co-relations of the gene expression levels of *E. coli* at 5 min and 1 h after shock with ZnO nanoparticles and ZnCl$_2$

nanoparticle–microbe interaction. Preliminary data analysis has shown a rapid differential expression of a large subset of genes with ZnO exposure. Among genes that were up-regulated *sodC*, a superoxide dismutase (Cu, Zn) gene, *ynfE*, *ynfF*, and oxidoreductase subunit genes have been previously associated with oxidative stress responses to ZnO nanoparticles [14]. Also of interest is the fact that many other genes are more significantly regulated with ZnO exposure. It will be interesting to see how co-regulated genes contribute to the stress response of the microbes. In addition, RNA profile of cells exposed to ZnO and ZnCl$_2$ is different suggesting exclusive toxicity of NPs to the microbes and confirming the hypothesis that the nanoparticle themselves set off a stress response in microbes. These results provide a better opportunity to study and compare genetic response and toxicology mechanisms involved with ZnO nanoparticles.

3.5 Summary

Release of engineered nanomaterials and/or nano-based products in the form of additives highlights one to gain knowledge on the interactions between these nanomaterials and biological systems. The current chapter summarizes the potential bactericidal toxicity of zinc oxide nanoparticles on the Gram-negative bacterium, *E. col,i* and Gram-positive bacterium, *B. subtilis*. The significance of utilizing thoroughly characterized nanomaterials and their synthesis achieved without the involvement of hazardous solvents, and chemicals have been emphasized. Materials usually from commercial sources may retain unknown surfactants or additives that can infer toxicity interpretations. Hydrothermally prepared small zinc oxide nanorods showed growth inhibition toward both the bacteria as a function of nanoparticle concentration. Additionally, the genome stress response of *E. coli*, in particular, toward ZnO nanoparticles, is outlined to assess the bacterial cell response.

References

1. D.A. Pelletier, A.K. Suresh, G.A. Holton, C.K. McKeown, W. Wang, B. Gu, N.P. Mortensen, D.P. Allison, D.C. Joy, M.R. Allison, S.D. Brown, T.J. Phelps, M.J. Doktycz, Appl. Environ. Microbiol. **76**, 7981–7989 (2010)
2. O.C. Farokhzad, J.J. Cheng, B.A. Teply, I. Sherifi, S. Jon, P.W. Kantoff, J.P. Richie, R. Langer, Proc. Natl. Acad. Sci. U.S.A. **103**, 6315–6320 (2006)
3. J.R. Morones, J.L. Elechiguerra, A. Camacho, K. Holt, J.B. Kouri, J.T. Ramirez, M.J. Yacaman, Nanotechnology **16**, 2346–2353 (2005)
4. M.H. Zhao, Z.L. Wang, S.X. Mao, Nano Lett. **4**, 587–590 (2004)
5. L.K. Adams, D.Y. Lyon, P.J.J. Alvarez, Water Res. **40**, 3527–3532 (2006)
6. E.G. Comini, G. Faglia, G. Sberveglieri, Z.W. Pan, Z.L. Wang, Appl. Phys. Lett. **81**, 1869–1871 (2002)
7. N. Jones, B. Ray, K.T. Ranjit, A.C. Manna, FEMS Microbiol. Lett. **279**, 71–76 (2008)
8. S.J. Klaine, P.J.J. Alvarez, G.E. Batley, T.F. Fernandes, R.D. Handy, D.Y. Lyon, S. Mahendra, M.J. McLaughlin, J.R. Lead, Environ. Toxicol. Chem. **27**, 1825–1851 (2008)
9. K.A.D. Guzman, M.R. Taylor, J.F. Banfield, Environ. Sci. Technol. **40**, 1401–1407 (2006)
10. C. Kirchner, T. Liedl, S. Kudera, T. Pellegrino, A.M. Javier, H.E. Gaub, S. Stolzle, N. Fertig, W.J. Parak, Nano Lett. **5**, 331–338 (2005)
11. L. Qi, Z. Xu, X. Jiang, C. Hu, X. Zou, Carbohydr. Res. **339**, 2693–2700 (2004)
12. A. Nel, T. Xia, L. Madler, N. Li, Science **311**, 622–627 (2006)
13. G.G. Xiao, M. Wang, N. Li, J.A. Loo, A.E. Nel, J. Biol. Chem. **278**(50), 50781–50790 (2003)
14. X. Tien, M. Kovochich, M. Liong, L. Mädler, B. Gilbert, H. Shi, J.I. Yeh, J.I. Zink, A.E. Nel, ACS Nano **2**(10), 2121–2134 (2008)

Chapter 4
Influence of Surface Coatings on the Bactericidal Activity of Nanoparticles

Abstract The existing literature on the influence of various surface coatings of nanoparticles in dictating bactericidal toxicity will be outlined. Various surface stabilizing agents on the toxicity of nanoparticles will be discussed. How engineered nanoparticles are incorporated in various surface coatings (chemical or biological) during their synthesis, along with details on the various physical characterization techniques including zeta potential, Fourier transform infrared spectroscopy (FTIR), and X-ray photoelectron spectroscopy (XPS), will be described. Finally, comparative studies on the effects of various surface-coated nanoparticles on the toxicity of bacteria will be discussed.

Keywords Surface · Coatings · Biocompatibility · Differential toxicity · Stability · Interactions

4.1 Surface Coatings or Stabilizing Agents

Nanoparticles usually have a primarily core material, which is then encapped by a shell or cap that acts as a stabilizing agent either to attain the overall stability of the nanomaterials stability and/or biocompatibility and/or reactivity. Surface stabilization is considered as one of the most important parameter, which not only can determine the fate of a nanomaterial in the environment but also the fate of its interaction with biological materials or cells, because this determines the primary mode of contact. Surface coating can affect the surface charge of the material that in turn can affect the affinity of the material to the cell surface. Depending on the surface coatings, the nanomaterial attains either a range of net negative (low to high) to positive charge (low to high). However, surface coating is not solely the one that determines the surface charge; this also depends on the primary core. For example, Wang et al., while assessing the antibacterial activity of a series of quantum dots including CdSe, CdTe, and ZnS–AgInS$_2$

© The Author(s) 2015
A.K. Suresh, *Co-Relating Metallic Nanoparticle Characteristics and Bacterial Toxicity*, SpringerBriefs in Biometals,
DOI 10.1007/978-3-319-16796-1_4

against luminous bacterium, *P. phosphoreum,* suggested that the differences in the overall surface charge induced by the surface coatings were responsible for bactericidal properties [1]. The authors opined that the phototoxicity generated either by using natural sunlight or artificial high-intensity lamps cause the direct release of metal ions that cause the toxicity. Feris et al., while studying the toxic effects of diethylene glycol (DEG), functionalized small ZnO nanoparticles with overall positive charge with a zeta potential of $+38.7 \pm 1.8$ mV on *P. aeruginosa* indicated that the cells due to their charged outer membrane surfaces are more susceptible to the particles [2]. The authors concluded that controlling the electrostatic attractions between the nanoparticles and the bacterium might permit the modulation of the bactericidal action. In another investigation, similar nanoparticle when treated with two different strains of pathogenic bacterium, *S. agalactiae* and *S. auereus,* showed dose-dependent cellular internalization (ref). Joshi et al., studying the influence of surface coatings of ZnO quantum dots of 3–5 nm size distributions, suggested that acetate adsorbed quantum dots were more toxic than nitrate adsorbed ones on E. coli with MIC of 2.5 and 6 mM, respectively [3]. Similarly, Brayner et al., evaluated the toxicological impacts of ultra-fine ZnO nanoparticles on *E. coli* in the presence of various surface stabilizers such as tri-n-octylphosphine oxide, sodium dodecyl sulfate (SDS), polyoxyethylene stearyl ether, and bovine serum albumin imparting various size and shape distributions and surface charges [4]. SDS-capped ZnO nanoparticles were found to be most toxic with 15 % inhibition and TOPO and Brij-76 a positive effect that may promote bacterial growth after their metabolism. Liong et al., while studying the interaction of silver nanoparticles encapsulated in mesoporous silica nanoparticles (Ag@MESs) (anionic, with a zeta potential of -22 mV) and Ag@MESs coated with polyethyleneimine (cationic with zeta potential of $+82$ mV) on *E.coli* and *B. anthracis,* suggested at concentrations of 100 μg/mL both types of particles completely inhibited bacterial growth, however, at 50 μg/mL, the delay of bacterial growth was observed only for negatively charged Ag@MESs [5]. Recently, Wigginton et al., evaluated the protein–nanoparticle interactions based on proteomic studies to understand the ecotoxicological impacts on model bacterium *E. coli* [6]. A number of proteins from *E. coli* were identified that specifically bind to bare carbonated Ag nanoparticles among which tryptophanase had the highest affinity despite its low abundance. The strong binding of Ag nanoparticles to TNase might have resulted in the significant reduction of enzyme activity leading to impaired metabolism and cell death. Very recently, Badawy et al., suggested that surface charge is the most important factor that governs the toxicity of silver nanoparticles. By using four different surface coating scenarios ranging from highly negative to highly positive suggested that the more negative particles were least toxic and the positively charged particles were most toxic. It appeared to the authors that the mechanism of toxicity might involve a combination of physical and chemical interactions and once the electrostatic barrier is overcome, the particle interacts with cell membrane causing physical damage by forming pits leading to cell death.

Additionally, some types of coatings can make engineered nanomaterials non-toxic; Aruguete et al., while evaluating the interactions of polymer encapsulated CdSe–CdS core/shell nanocrystallites observed that the quantum dots were not interacting with bacteria at all and were non-bactericidal [7]. Though using another form of nanomaterial (Ag), similar surface coatings dependent on differential toxicities were also made by Suresh et al., while evaluating the comparative toxicities of different surface-encapped silver nanoparticles (colloidal-Ag with no surface coat and a zeta potential of -42 ± 5 mV; biogenic-Ag with protein/peptide surface coat and a zeta potential of -12 ± 2 mV, and oleate-Ag with oleate as a surface coat and a zeta potential of -45 ± 4 mV) [8]. The authors suggested that biogenic-Ag was most toxic than the colloidal-Ag toward Gram-negative (*E. coli, S. oneidensis*) and Gram-positive (*B. subtilis*) bacterial strains, whereas oleate-stabilized silver nanoparticles, despite being the smallest ones, were not toxic to any of the strains that were evaluated. Overall, these discussed literatures highlight the influence of surface charge, as a measure of zeta potential, of engineered nanocrystallites in dictating the bactericidal toxicities. The more positively charged nanoparticles were found to be the most toxic. Earlier reports suggest that neutral and negatively charged particles adsorb to a lesser extent on negatively charged cell membranes and consequently show lower levels of internalization when compared to positively charged particles [9]. However, the exact mechanism of the effect of surface coating-dependent interactions, their specific retentions, and the genetic responses involved still needs some investigations. Apart from the above-mentioned, several other factors can also significantly influence the toxic response of engineered nanomaterials toward bacteria such as growth medium and medium components, and the presence or absence of additives.

Growth medium and medium components do influence the toxic response of nanoparticles, for instance, based on our previous results while assessing the influence of engineered cerium oxide nanoparticles on multiple bacterial strains, we observed growth inhibition toward *E. coli* and *B. subtilis* only in minimal medium (as a function of the nanoparticle size) and not Luria–Bertani medium. In contrast, *S. oneidensis* growth was not at all inhibited by the cerium oxide nanoparticles [10], indicating that the observed size-dependent response may result from size-dependent characteristics of the cerium oxide nanoparticles and/or metabolic characteristics of the different organisms. Similarly, Li et al., suggested that water chemistry is a major factor in regulating the toxicity mechanism of ZnO nanoparticles to *E. coli* [11]. In their investigation on the influence of ZnO in the presence of five different mediums including ultra-pure water, 0.85 % sodium chloride, phosphate-buffered saline, minimal Davis, and Luria–Bertani revealed that the toxicity decreases as follows: ultra-pure water > sodium chloride > minimal Davis > Luria–Bertani > phosphate-buffered saline. The authors opined that the generation of precipitates involving PBS and zinc complexes involving minimal Davis and Luria–Bertani medium drastically decreased the concentration of Zn^{2+} ions thereby lowed toxicities in these mediums [11].

Likewise, the presence or absence of additives can also influence nanoparticle-mediated toxicity, for example, Stoimenov et al., in their investigation on the interaction of reactive magnesium oxide nanoparticles and halogen adducts with diverse bacterial strains as well as spores, illustrated considerable changes in the cell membrane upon treatment resulting in cell death [12]. The authors proposed that abrasiveness, basic character, electrostatic attractions, and oxidizing power (due to the presence of halogen) combine to promote the biocidal properties. Likewise, Li et al., investigated the effects of iron-doped ZnO nanoparticles on *B. subtilis* and *E. coli* and suggested that *E. coli* are more resistant than *B. subtilis* with the IC50 values of 15–43 and 0.3–0.5 mg/L, respectively [13]. Additionally, the authors assessed the influence of acid additives and suggested that tannic acid decreased the toxicity of ZnO nanoparticles over humic, fulvic, and alginic. The authors opined that the tannic acid complexes the most free Zn^{2+} ions, thereby reducing their bioavailability. Rincon et al., while evaluating the effects of multiple chemical parameters such as pH (over a range of 4–9), inorganic ions ($HCO3-$, HPO_4^{2-}, Cl^-, NO^{3-}, and SO_4^{2-}), organic matter (dihydroxybenzenes, hydroquinones, catechol, and resorcinol), and hydrogen peroxide on the photocatalytic inactivation rate of TiO_2 nanoparticles on *E. coli*, observed parameter-based differential effects. The authors noted that pH did not affect while hydrogen peroxide had positive impact on *E. coli* inactivation rate. Similarly, the presence of different additives in the form of inorganic ions showed weak to strong influence on the photocatalytic bactericidal effects. Finally, in the case of when organic matter was used, the organic matter itself was degraded and did not have any significant effect on the bacterial inactivation [14].

4.1.1 Various Chemical and Biological Stabilizing Agents

The implementation of various surface-capping or surface coating agents allows the stabilization of the nanoparticles including their synthesis. Nanoparticles are often coated with surface-capping agents composed of either inorganic materials or polymeric materials. Chemically, these stabilizing molecules are called surfactants and the most commonly used surfactants are the detergents: SDS, sodium oleate, dodecylamine, sodium carboxymethylcellulose (CMC), betyltrimethylammonium bromide (BTAB), cetyltrimethylammonium bromide (CTAB), mercaptopropionic acid (MPA), mercaptoundecanoic acid (MUA), and beta-mercaptoethanol, Whereas polymeric coatings include the use of synthetic polymers such as polyvinyl alcohol (PVA), polyvinyl pyrrolidone (PVP), polylactic-co-glycolic acid (PLGA), polyethylene glycol (PEG), and polyethylene-co-vinyl acetate, the scarcity or shortage of the polymers prompts researches to always look for new surfactants from natural sources, inspired by the nature.

Biomimetics is well known naturally occurring phenomenon that mediates the formation of a wide range of inorganic materials including bone, shells, lenses, and even nanoparticles (magnetic particles formed by the magneto tactic bacteria). Biological systems have adapted to extreme physical, immunological, defense, and weather conditions by implementing diverse dynamic biomimetic approach to the proper functioning of the system, either it may be an enzyme catalysis that mediates the formation of growing crystal structures (shells). As a consequence, evolution and natural selection have resulted in organisms capable of synthesizing materials with controlled compositions, microcrystalline shapes, and long-range organization. Inspired from nature, researchers are putting efforts to rebuild or re-architect biomimetic for the benefit of mankind and nature at large. Several biological materials to list a few antibodies, proteins, nucleic acids, nucleotides, polysaccharides, glycoproteins, metabolites, polymers (gelatin, dextran, and chitosan), etc., have been used as synthesis and stabilizing agents to produce diverse forms of nanoparticles. The advantages of using biomaterials are, as these are derived from the biological systems, which function at moderate conditions, ambient pH, temperature and pressures, and aqueous suspensions. In addition, these processes are more ecologically benign to fabricate various inorganic nanomaterials; moreover, the use of selective ligand or precursor moieties can facilitate the assembly of nanoscale materials into more complex functional assemblies and smart devices.

4.2 Assessing the Incorporated Surface Coating on the Nanoparticles Surface

Several characterization techniques are used to evaluate the encapped surface-stabilizing agent associated with the nanoparticles. The most commonly used techniques are described below.

4.2.1 Zeta Potential

Zeta potential is used to measure the magnitude of electrostatic or charge repulsion or attractions between the nanoparticles and is one of the fundamental parameters that are known to determine the overall stability of the nanoparticles. Scientifically, zeta potential is a term for electrokinetic potential in colloidal dispersions, whereas theoretically, zeta potential is the potential difference between the dispersion medium and the stationary layer of fluid attached to the dispersed particle. The zeta potential is caused by the net electrical charge contained within the region bounded by the slipping plane, which also depends on the location of that plane. Thus, it is widely used for quantification of the magnitude of the charge. The magnitude of the zeta potential indicates the degree of electrostatic repulsion between adjacent, similarly charged particles in dispersion. For molecules and particles that are small enough, a high zeta potential will confer stability, i.e., the solution or dispersion will resist aggregation. Zeta potential measurement can provide detailed insight on various aspects of nanoparticles such as

Table 4.1 Differential surface charge observed for similar nanoparticles (Ag) coated with different surface coatings as a measure of zeta potential

Similar nanoparticle @ different surface coating	Zeta potential (mV)
Ag@poly(diallyldimethylammonium)	+45 ± 3
Ag@protein coat	−12 ± 2
Ag@oleate	−45 ± 4
Ag@uncoated	+42 ± 5

nanoparticle dispersion, nanoparticle aggregation or nanoparticle flocculation, and nanoparticle overall surface charge. The measurement of zeta potential has tremendous applications including ceramics, pharmaceuticals, biomedicine, mineral processing, electronics, and wastewater treatment.

For biological experimentation, zeta potential measurements were performed on a Brookhaven 90 Plus/BI-MAS Instrument (Brookhaven Instruments, New York). With respect to nanoparticle-stabilizing agents, zeta potential is used as a measure of the overall surface charge of the particles. If similar form of nanoparticles is coated with different surface-stabilizing agents, the overall net charge will be different for the different coatings used. For example, Table 4.1 illustrates the zeta potential measurements of silver nanoparticles coated with four different stabilizing agents: PDADMAC-Ag containing a poly(diallyldimethylammonium) surface coat, biogenic-Ag containing a protein or peptide surface coat, oleate-Ag with oleate surface coat, and uncoated-Ag with no surface coating. As can be made out from the table, the PDADMAC-Ag nanoparticles were positively charged with a zeta potential of +45 ± 5 mV, biogenic-Ag nanoparticles were the least negatively charged with a zeta potential of −12 ± 2 mV, uncoated-Ag nanoparticles had a zeta potential of −42 ± 5 mV, and oleate-Ag nanoparticles had a zeta potential of −45 ± 4 mV [9]. Clearly indicating that zeta potential measurements can be used as a means to assess the different surface or stabilizing agents associated with the nanoparticles based on their overall surface charge.

4.2.2 Fourier Transform Infrared Spectroscopy

Fourier transform infrared spectroscopy (FTIR) is mostly useful for identifying molecules that are either organic or inorganic. It can also be utilized to quantitate some components of an unknown mixture and for the analysis of various forms of solids, liquids, and gasses. FTIR involves the vibration of chemical bonds of a molecule at different frequencies depending on the elements and types of bonds. After absorbing electromagnetic radiation, the bond frequency increases leading to transition between ground and excited states. These absorption frequencies represent excitations of vibrations of the chemical bonds and are specific to the type of bond and the group of atoms involved in the vibration. The energy of these frequencies is in the infrared region (4000–400 cm^{-1}) of the electromagnetic

Fig. 4.1 Fourier transform infrared spectroscopy of the silver nanoparticles surface functionalized with protein surface coating

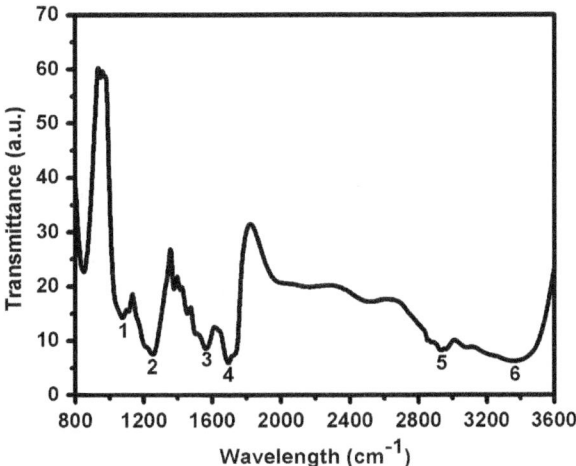

spectrum. The term Fourier transform refers to a recent development wherein the data obtained in the form of an interference pattern are converted to an infrared absorption spectrum, similar to that of a molecular "fingerprint."

To determine the nature of stabilizing agent surrounding the nanoparticles, FTIR measurement can be applied. FTIR analysis of the liquid or dried samples in KBr pellet is performed on a Thermo Nicolet model 6700 spectrophotometer in a diffuse transmittance mode at a resolution of 4 cm^{-1}. The FTIR analysis generally reveals the presence of strong bands centered at various frequencies depending on the surface coating. For example, Fig. 4.1 demonstrates the FTIR spectroscopy for silver nanoparticles that are coated with a protein surface coating. Depending on the vibration bands generated from the protein coat vibration, peaks centered at 1080, 1365, 1640, 1767, 2425, 2913, and 3392 cm^{-1} can be seen. Data analysis of the various bands reveals their functional groups. The vibrational band at 1080 cm^{-1} corresponds to alcoholic and carbonyl groups. Peak at 1365 cm^{-1} corresponds to amide III functional group. Band at 1640 cm^{-1} is due to the presence of carbonyl (–C–O–C– or –C–O) stretch and –N H– stretch in amide linkage, and vibrational peak at 2913 cm^{-1} indicates the presence of C–H group in C–CH$_3$ compounds. Peak at 3392 cm^{-1} is the characteristic of the hydroxyl functional (–OH) group in alcohols and phenolic compounds. Overall, these peaks clearly suggest the presence of protein or a peptide on the surface that likely appears to be acting as a capping molecule.

4.2.3 X-ray Photoelectron Spectroscopy

X-ray photoelectron spectroscopy (XPS) also known as electron spectroscopy for chemical analysis (ESCA) is the most widely used surface analytical technique

Table 4.2 Chemical surface composition (%) of various elements detected in the silver sulfide nanoparticles coated with protein surface coating

Element	Ag3d	C1s	Ca2p	N1s	Na1s	O1s	P2p	S2p	Si2p
Ag2S@Protein	1.0	91.0	0.0	0.6	traceable	5.9	0.0	1.1	0.5

because of its relative simplicity in use and data interpretation. XPS is widely used for probing the electronic structures of atoms, molecules, and condensed matter. When an X-ray photon of energy $(h\nu)$ is incident on a solid matter, the kinetic energy (E_k) and the binding energy (E_b) of the ejected photoelectrons can be related as $E_k = h\nu - E_b$. This kinetic energy distribution of the photoelectrons forms a series of discrete bands, which symbolizes the electronic structure of the sample. The core-level binding energies of all the elements (except H and He) in different oxidation states are unique which provides information about the chemical composition of the sample. However, to account for the multiplet splitting and satellites accompanying the photoemission peaks, the photoelectron spectra should be interpreted in terms of many-electron states of the final ionized state of the sample, rather than the occupied one-electron state of the neutral species.

XPS is a well-known technique to evaluate the orbital orientation as well as elemental composition of the materials. Well will demonstrate the applicability of XPS to assess nanoparticle surface coatings illustrating an example from our own work where we made silver sulfide nanoparticles coated with protein [15]. In a typical XPS analysis, a survey spectral sweep of the samples was collected over a range of binding energies (0–1400 eV) and that will show the presence on individual elements, and for the biogenic silver sulfide, the presence of Ag, S, C, O, Ca, N, P, and Na was detected. The surface composition (%) of various elements detected for the biogenic silver sulfide nanoparticles coated with protein is given in Table 4.2. For the purified silver sulfide nanoparticle sample, the Ag3$d_{5/2}$ and S2p peaks were observed at 368 and 162 eV, respectively, corresponding with the predicted values for silver attached to sulfide [15]. The Ag3d spectrum could be decomposed into a single spin–orbit pair Ag3$d_{5/2}$ and Ag3$d_{3/2}$, with a spin–orbit splitting of ~6 eV and with peaks observed at binding energies of 368.1 and 374.1 eV, respectively [15]. The spectrum for S2p showed peaks at 161.4 and 162.1 cV that could be assigned to the binding energies of S2$p_{3/2}$ and S2$p_{1/2}$, correspondingly, with a single pair spin–orbit splitting of ~0.7 eV [15]. The sulfide peaks at 162 eV and a sulfate species at 168 eV were observed in the S2p region. These data clearly suggest that the particles were encapped by a proteinaceous material.

4.3 Summary

Applications of nanoparticles in various consumer products require understanding of their surface-stabilizing agents. Factors such as the surface charge that lead to differential binding, differential interaction, and differential aggregation potential, which are influenced by the surface coatings, are a major determinant factor in dictating the potential toxicity and bacterial interactions of nanoparticles. This chapter summarizes the roles of various biological and surface coatings inducing bactericidal activity. The various physical characterization or analytical techniques that can shed information on the details of the surface coatings that is associated with the particles are summarized. These insights can be exploited for defining nanoparticles with specific surface coatings eliciting toxic responses.

References

1. L.L. Wang, H.Z. Zheng, Y.J. Long, M. Gao, J.Y. Hao, J. Du, X.J. Mao, D.B. Zhou, J. Hazard. Mater. **177**, 1134–1137 (2010)
2. K. Feris, C. Otto, J. Tinker, D. Wingett, A. Punnoose, A. Thurber, M. Kongara, M. Sabetian, B. Quinn, C. Hanna, D. Pink, Langmuir **26**, 4429–4436 (2010)
3. P. Joshi, S. Chakraborti, P. Chakrabarti, D. Haranath, V. Shanker, Z.A. Ansari, S.P. Singh, V. Gupta, J. Nanosci. Nanotechnol. **9**, 6427–6433 (2009)
4. R. Brayner, R. Ferrari-Iliou, N. Brivois, S. Djediat, M.F. Benedetti, F. Fievet, Nano Lett. **6**, 866–870 (2006)
5. M. Liong, B. France, K.A. Bradley, J.I. Zink, Adv. Mater. **21**, 1684–1689 (2009)
6. N.S. Wigginton, A. De Titta, F. Piccapietra, J. Dobias, V.J. Nesatty, M.J.F. Suter, R. Bernier-Latmani, Environ. Sci. Technol. **44**, 2163–2168 (2010)
7. D.M. Aruguete, J.S. Guest, W.W. Yu, N.G. Love, M.F. Hochella, Environ. Chem. **7**, 28–35 (2010)
8. A.K. Suresh, D.A. Pelletier, W. Wang, J.-W. Moon, B. Gu, N.P. Mortensen, D.P. Allison, D.C. Joy, T.J. Phelps, M.J. Doktycz, Environ. Sci. Technol. **44**, 5210–5215 (2010)
9. A.K. Suresh, D.A. Pelletier, W. Wang, J.L. Morrell-Falvey, B. Gu, M.J. Doktycz, Langmuir **28**, 2727–2735 (2012)
10. D.A. Pelletier, A.K. Suresh, G.A. Holton, C.K. McKeown, W. Wang, B. Gu, N.P. Mortensen, D.P. Allison, D.C. Joy, M.R. Allison, S.D. Brown, T.J. Phelps, M.J. Doktycz, Appl. Environ. Microbiol. **76**, 7981–7989 (2010)
11. M. Li, L.Z. Zhu, D.H. Lin, Environ. Sci. Technol. **45**, 1977–1983 (2011)
12. P.K. Stoimenov, R.L. Klinger, G.L. Marchin, K.J. Klabunde, Langmuir **18**, 6679–6686 (2002)
13. M.H. Li, S. Pokhrel, X. Jin, L. Madler, R. Damoiseaux, E.M.V. Hoek, Environ. Sci. Technol. **45**, 755–761 (2011)
14. A.G. Rincon, C. Pulgarin, Appl. Catalysis B Environ. **51**, 283–302 (2004)
15. A.K. Suresh, J. Mitchel Doktycz, W. Wang, J.W. Moon, B. Gu, M. Harry Meyer III, K. Dale Hensley, P. David Allison, J. Tommy Phelps, A. Dale Pelletier, Acta Biomater. **7**, 4253–4258 (2011)

GPSR Compliance

*The European Union's (EU) General Product Safety Regulation (GPSR)
is a set of rules that requires consumer products to be safe and our
obligations to ensure this.*

*If you have any concerns about our products, you can contact us on
ProductSafety@springernature.com*

In case Publisher is established outside the EU, the EU authorized
representative is:

Springer Nature Customer Service Center GmbH
Europaplatz 3
69115 Heidelberg, Germany

Batch number: 09478958

Printed by Printforce, the Netherlands